ID0984934

Springer Series on
SIGNALS AND COMMUNICATION TECHNOLOGY

SIGNALS AND COMMUNICATION TECHNOLOGY

continued after index

Marc Engels
Frederik Petré

Broadband Fixed Wireless Access

A System Perspective

 Springer

Marc Engels
Frederik Petré

Broadband Fixed Wireless Access: A System Perspective

Library of Congress Control Number: 2006926213

ISBN 0-387-33956-6   e-ISBN 0-387-34593-0
ISBN 978-0387-34593-2

Printed on acid-free paper.

Printed in the United States of America.

9 8 7 6 5 4 3 2 1

springer com

Contents

List of Figures

List of Tables

Preface

With the advent of the IEEE 802.16 standard, Broadband Fixed Wireless Access (BFWA) is becoming a new hype for high-speed internet access for residential customers and small and medium enterprises (SMEs). Up till now, a large variety of incompatible wireless solutions were promoted for BFWA. Moreover, these solutions only worked under restricted operating conditions and hence had a high operational cost for installation and maintenance. Both shortcomings are eliminated by the new standard.

Recognizing the importance of this standard, we started in 2001 a spin-off company of IMEC, LoraNet, with funding of the IMEC Incubation Fund. LoraNet had the mission to become a leading provider of BFWA technologies. LoraNet could capitalize on the IMEC know-how on Wireless Local Area Network (WLAN) systems and actively developed techniques to reduce the total cost of ownership of a BFWA solution for the operators.

To this end, we participated in the development of the IEEE 802.16 standard and its European counterpart, HIPERMAN. LoraNet also executed an SME development project that was funded by the Institute for the Promotion of Innovation by Science and Technology in Flanders (IWT). In addition, LoraNet participated in the European IST-STRIKE project, which investigates multi-antenna techniques for BFWA as well as the bridging between BFWA and WLAN.

Due to the lack of venture capital LoraNet had to be closed down in mid 2003. Nevertheless, the interest in BFWA is still growing and we did not want that the experience that was build up in LoraNet, got lost. Therefore we bundled our experience in a tutorial for the 2004 International Conference on Communications (ICC) in Paris. The success of this tutorial led us to the idea to reach a wider audience by extending the tutorial material into a manuscript.

The result of this effort is in front of you. We hope that you enjoy reading it and that it is useful in your professional work.

Marc Engels

Contributors

Marc Engels is currently the general manager of Flanders' Mechatronics Technology Centre, a research centre that aims at increasing the intelligence of mechatronic systems. Research topics include intelligent sensors, communication, high-dynamic control and embedded software. Before, Marc Engels was the CTO of LoraNet, a start-up in the field of broadband wireless communication. Previously, he was the director of the wireless department at IMEC, focused on the implementation of telecommunication systems on a chip. For these systems, he overlooked research on the DSP processing, the mixed-signal RF front-end and the software protocols. He was also active in design methods and tools for implementing multi-disciplinary systems. Under his supervision, several systems have been realized, including a 54 Mbps WLAN terminal, a GPS-GLONASS receiver, a DECT-GSM dual mode phone, a cable modem, etc. Previously, Marc performed research at the Katholieke Universiteit Leuven, Belgium, Stanford University, CA, USA, and the Royal Military School, Brussels, Belgium. Marc Engels received the engineering degree (1988) and the Ph.D. (1993), both from the Katholieke Universiteit Leuven, Belgium. Marc Engels is a visiting professor of embedded system design at the Advanced Learning and Research Institute, University of Lugano, Switzerland.

Nadia Khaled was born in Rabat, Morocco, in 1977. She received the M. Sc. degree in electrical engineering from l' Ecole Nationale Supérieure d'Electrotechnique, d'Electronique, d'Informatique, d'Hydraulique et des Télécommunications (ENSEEIHT), Toulouse, France in 2000. On December 2005, she received her Ph.D. in electrical engineering from the Katholieke Universiteit Leuven (KULeuven), Leuven. From 2000 to 2005, she was with the Wireless Research Group of IMEC, Leuven. Since October 2005, she is a postdoctoral researcher at the Communication Theory Group at the Swiss Federal Institute of Technology Zurich (ETHZ), Switzerland. Her research interests lie in the area of signal processing for wireless communications, particularly MIMO techniques and transmit optimization schemes.

Frederik Petré is a Senior R&D Engineer and the Wireless Sensor System Architect at the Flanders' MECHATRONICS Technology Centre (FMTC), which is a new research centre, operating since October 2003, with the mission to establish a bridge between the academic and industrial know-how in mechatronics in Flanders, Belgium. Over there, he focuses on end-to-end system design and integration of mobile wireless sensor systems for industrial process monitoring and control applications. Before joining FMTC, Frederik was a Senior Scientist within the Wireless Research Group at the Interuniversity Micro-Electronics Centre (IMEC), investigating baseband signal processing algorithms and digital architectures for future generation wireless communication systems, including Third Generation (3G) and Fourth Generation (4G) broadband cellular networks and High-Throughput Wireless Local Area Networks (HT-WLANs). He received the M.S. degree (1997) and the Ph.D. (2003) in Electrical Engineering, both from the Katholieke Universiteit Leuven, Belgium. During the Fall of 1998, he spent 6 weeks as a visiting researcher at the Information Systems Laboratory (ISL), Stanford University, California, USA, working on OFDM-based powerline communications. Frederik is a member of the ProRISC Technical Program Committee and secretary of the IEEE Benelux Section on Communications and Vehicular Technology (CVT). In 2005, he served as a guest editor for the EURASIP Journal on Wireless Communications and Networking (JWCN), resulting in a special issue on *Reconfigurable Radio for Future Generation Wireless Systems*. From January 2004 till December 2005, he was a member of the Executive Board of the European 6th framework Network of Excellence in Wireless Communications (NEWCOM), and the leader of NEWCOM Project D on *Flexible Radio*.

Rafael Torres was born in Málaga, Spain, in 1961. He received his MS degree in Physics from the University of Granada (Spain) in 1986 and his Ph.D. from the Telecommunications Engineering School at the Polytechnic University of Madrid (UPM) in 1990. From 1986 to 1990 he was with the Radio Communication and Signal Processing Department of the UPM as a research assistant. During this time, he worked on numerical methods in electromagnetism, and its applications to design of passive microwave devices like radomes, circular polarizer, rotators and planar lenses. He became an associate professor in the Department of Communication Engineering of the University of Cantabria (Spain) in 1990. From this time to the present, he has participated in several projects about RCS computation, on board antennas analysis, electromagnetic compatibility, and radio-propagation. He is co-author of a book about the CG-FFT method, author of several chapters in different books, more than 20 papers, and about 80 conference contributions. He has been the leader of the group that has developed CINDOOR, Computer Tool for Planning and Design of Wireless Systems in Enclosed Spaces. His current research interests include radio-propagation for wireless and mobile communications, as well as the simulation and design of new wireless communications systems including MIMO.

Acknowledgements

This book was only possible with the help and support of many people. In the first place, I like to thank all the authors that contributed to the various chapters.

The material in the book is the result of an exciting experience with LoraNet. I am grateful to whole LoraNet team for the excellent atmosphere and the excellent results.

Finally, a word of thanks is due to my wife Els and my three daughters Heleen, Laura and Hanne for their patience and support.

Marc Engels

First, I would like to express my gratitude to Marc Engels for offering me the opportunity to be deeply involved in the exciting process of creating a book on BFWA. Furthermore, I would like to thank the other contributors, Nadia Khaled, and Rafael Torres, for their valuable input and contributions on some of the key topics. Finally, I would like to dedicate this book to my parents, Willem and Sylvia, for their unconditional love, support, and inspiration over the years.

Frederik Petré

Chapter 1

The Need for BFWA
Introduction

Marc Engels

1.1 BROADBAND FIXED WIRELESS ACCESS

Fixed wireless systems have a long history. Indeed, the first radio experiments by Marconi were based on fixed transmitters and receivers [1]. Since then, point-to-point microwave connections have been used for voice and data communication applications. Especially for backhaul networks of operators, point-to-point connections will continue to play an important role.

In the late '80s of the last century, narrowband fixed wireless point-to-multipoint systems started to be deployed for the provisioning of local telephone service, especially in developing countries. These systems were initially based on proprietary technology, but later on variants of cordless telephone standards, like DECT, PHS, and CT-2, became more popular. Recently, most authors advocated cellular standards, like GSM and GPRS, as the most cost-effective solution [2].

In an attempt to capitalize on the economies of scale in the delivery of services, several operators started in the late '90s to work on the bundling of voice service with data communication and television delivery [3]. This automatically led to the need for Broadband Fixed Wireless Access (BFWA) technology. BFWA tries to realize wireless communication with a user data rate exceeding 1 megabits per second (Mbps) between stationary users and base stations over distances of several kilometers. Initially, multiple proprietary technologies have been proposed for BFWA. However, these technologies failed due to their lack of interoperability, their high investment and operational costs, and their technological shortcomings. Recently, BFWA systems were standardized by the IEEE 802.16 Broadband Wireless

Access Working Group in the U.S. and by the ETSI BRAN committee in Europe. In addition, the WiMAX forum was established to promote and certify interoperable products based on the IEEE 802.16 standard.

The advent of these standards has created a renewed interest in BFWA, not only from operators and equipment manufacturers, but also in academic research institutes. Several analysts believe that BFWA will become a serious contender for alternative broadband access technologies, like digital subscriber line (DSL) and cable modems.

In this chapter, we discuss the market potential of the BFWA technology. First, we review the drivers that continuously increase the demand for broadband access. Next, we compare BFWA with its competing technologies. Based on its strong competitive edge, we continue with estimating the potential market size and identifying the critical factors for market success. Finally, we review the outline of the book.

1.2 THE QUEST FOR BROADBAND

About a decade ago, the telecommunication infrastructure was targeted towards fixed analogue telephony, with support for voice and narrow band data communication. In 10 years, the digitalization of the communication infrastructure, the support for broadband access at home, and the success of mobile phones have changed this situation completely.

End-to-end digitalization of the telecom network was achieved with the introduction of the Integrated Services Digital Network (ISDN) [4]. However, market success was limited due to the lack of interesting services. This changed dramatically with the fast take-up of the Internet by business and residential customers in the '90s.

The rapid adoption of internet, not only for business users but also for residential customers, is one of the major factors that continuously stimulate the demand for bandwidth. As shown in *Figure 1.1*, the number of worldwide internet users will have increased by more than an order of magnitude, from 138 million in 1998 to more than 1.5 billion in 2008 in only ten years time. Moreover, this trend does not show any sign of slowing down. More specifically, where originally the internet users were clustered in developed countries in North America and Europe, we currently observe a growing penetration in developing countries, like China and India. Several governments have recognized the importance of a broadband internet and access infrastructure, and, hence, are actively promoting it. South-Korea is a noteworthy example of active government support, which resulted in a pole position for almost all internet statistics.

Internet access is not limited to web browsing only. Especially for the residential users, a number of additional services are emerging. These range from relatively low rate services, like tele-shopping, to demanding multimedia applications like music and photo sharing. The user-to-user aspect of the latter applications also makes that the bandwidth demands become increasingly symmetrical. As a next step, multiple operators are considering the deployment of several variants of video- and audio-on-demand. This would allow them to offer a triple play service package, consisting of voice, data, and video services, over a single network.

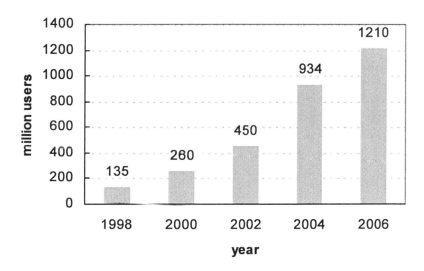

Figure 1.1. The evolution of internet users (source: clickz [5]).

Finally, we like to point out that a broadband pipe enables a different service deployment model. Where several services, like home automation and surveillance, traditionally require a home gateway, a broadband pipe makes it possible to provide these services in the network. Such approach could provide operators with an additional source of revenues.

The success of the Internet rapidly created a market for data pipes to individual users. The 128 kbps of ISDN was no longer considered sufficient. Hence, Asymmetrical Digital Subscriber Line (ADSL) [6] and cable modem [7] technology were developed to increase the data rate to several hundreds of kbps, with a theoretical maximum of 10 Mbps. Very high speed Digital Subscriber Line (VDSL) technology [8], which will further increase this data rate up to 52 Mbps, is currently being investigated for market introduction.

The spectacular growth of internet, also in developing countries, and the interest of operators in developed countries for triple play service packages, that comprise voice, data, and video, stimulate the demand for broadband access. Although BFWA has currently a marginal market share, it is expected that, from 2006 onwards, BFWA will become an important contender for digital subscriber line (DSL) and cable modem technologies.

1.3 COMPETING TECHNOLOGIES

Where traditionally the most popular internet access technology was the narrowband telephone modem, DSL and cable modem technologies have experienced a remarkable growth in recent years. According to the figures of the Organization of Economic Cooperation and Development (OECD), as shown in *Figure 1.2*, Korea has more than 20 broadband access users per 100 inhabitants. Several European countries followed the strategy of actively promoting competition on the access network, resulting in broadband penetrations between 5% and 15%. Nevertheless, a comparison with South Korea learns that there is still a large growth potential.

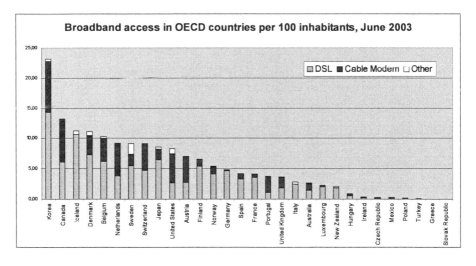

Figure 1.2. Broadband access penetration (source: OECD [9]).

1.3.1 Digital Subscriber Line (DSL)

Asymmetrical digital subscriber line (ADSL) technology upgrades the plain old telephone system (POTS) for broadband access [10]. It allows for simultaneous voice communication and transmission of digital data on a single copper pair. The ADSL data is transmitted at frequencies above the existing telephony band. At the customer premises, a splitter multiplexes the POTS and the ADSL signals, while preventing mutual interference of both signals. In order to combine the signals from (upstream) and towards (downstream) the customer on a single telephone line, the ADSL standard allows for two variants. The first one has a downstream signal overlapping the upstream signal, and, hence, requires echo cancellation (EC) to separate both signals. The second one uses frequency division duplexing (FDD) with separate frequency bands for the up- and downstream.

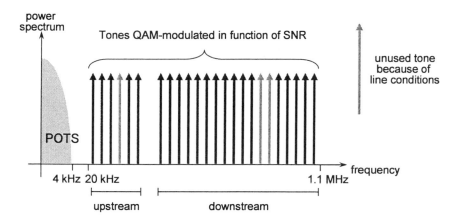

Figure 1.3. ADSL FDD DMT modulation.

The dominant modulation scheme for ADSL is Discrete MultiTone (DMT). In DMT, the frequency band is divided in subchannels or tones: 32 for the upstream and 256 for the downstream. Each of these subchannels is transmitting a quadrature amplitude modulation (QAM) signal. The QAM modulation order for each subchannel is independently chosen, depending on the signal-to-noise-ratio (SNR) for this subchannel. The ADSL DMT modulation is schematically illustrated in *Figure 1.3*.

On a good copper line, the downstream ADSL data rate exceeds 8 Mbps and the upstream data rate approaches 1 Mbps. However, ADSL is unable to support every user with high bandwidths. Indeed, *Figure 1.4* shows that the distance of customers to a central office varies largely from country to country [11]. This variation is caused by the difference in network architecture. As the data rate of a digital subscriber line decreases with the distance, the achievable data rate shows a similar variation. For example, in the United States, with very long local loops, only 35% of the potential users are currently capable of receiving more than 1.5 Mbps. In Europe, ADSL has a much brighter future with over 70% of potential users that have broadband capabilities.

The copper loop lengths could be reduced to a few hundred meters by deploying fibers to the street cabinets or even to the building. This would enable very high speed digital subscriber lines (VDSL) with a downstream data rate of up to 52 Mbps, but at the cost of a considerable investment.

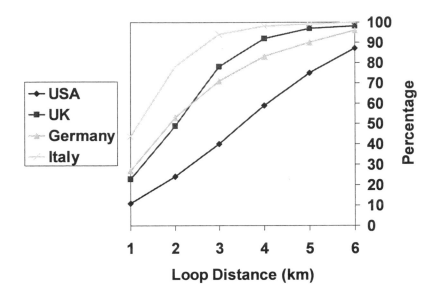

Figure 1.4. Loop length distribution (according to [11]).

1.3.2 Cable

Cable modems on community antenna television or cable television (CATV) networks are an attractive alternative to digital subscriber lines [12]. Their superior hybrid fiber coax (HFC) physical network, consisting of optical fiber and shielded coaxial cables and amplifiers with bandwidths up to 860 MHz, makes that almost unlimited data rates can be realized. Moreover, the cable network is naturally suited for video broadcasting.

Most cable modems are compatible with the Data over Cable System Interface Specification (DOCSIS) for bidirectional data communication over the HFC medium. DOCSIS employs unused portions of the HFC spectrum between 5 and 42 MHz for upstream transmissions and between 550 and 750 MHz for downstream transmissions. These frequency bands are further divided into individual channels. For downstream transmissions, 6 MHz channels are used in the U.S, and 8 MHz in Europe. These channels are QAM modulated with constellations of up to 256 points. For upstream transmissions, the allocated spectrum is divided into individual channels, with bandwidths ranging from 200 to 3200 kHz. Both quaternary phase shift keying (QPSK) and 16-QAM are permitted in these upstream channels. The aggregate bandwidth over all upstream channels is typically around 30 Mbps, while the aggregate downstream bandwidth is an order of magnitude larger.

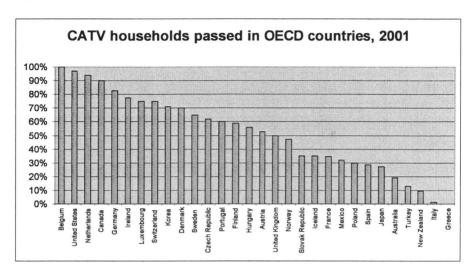

Figure 1.5. Presence of CATV network (source: OECD [9]).

The main shortcoming of this technology is the limited presence of the cable network in several major markets, as shown in Figure 1.5. In Italy, for instance, less than 2% of the homes are passed by a cable network. In countries with a high presence, like Belgium, the United States, or Germany, cable modems are successfully deployed. However, the transformation of a uni-directional broadcasting network into a bi-directional communication network requires a huge investment. Indeed, the traditional branch-and-tree structure of the CATV network has to be partitioned, and fiber optic links have to be run from the head end towards fiber network units within the old CATV plant. In Flanders (the northern part of Belgium), for instance, a roll-out plan of more than 5 years was needed to enable data communication over the CATV network.

1.3.3 Emerging alternatives

Up till now, we may conclude that there is a rising demand for broadband access and that digital subscriber line and cable modem technologies can only provide a partial answer to address this demand. As a consequence, there is a need for a third broadband access technology to fill the gap.

Several alternatives have been proposed: optical fiber, free space optics, satellites, and power line communication.

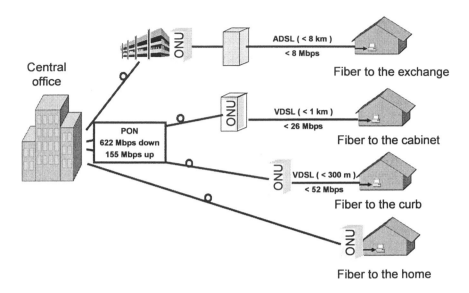

Figure 1.6. Hybrid fiber-copper architectures.

Optical fiber could provide virtually unlimited data rates but is a very costly solution, especially when trenches have to be dug into existing habitations to install the fiber [14]. Costs of up to €100 per meter are quoted for city centers. As a consequence, several intermediate solutions have been proposed to combine fiber with coax or copper cables. Depending on the place where the optical network unit (ONU) that converts the optical signals into electrical ones, is placed in the network, a distinction is made between fiber-to-the-home, fiber-to-the-curb, fiber-to-the-cabinet, or fiber-to-the-exchange [13]. A schematic representation of these different architectures can be found in *Figure 1.6*. The closer the ONU is placed to the central office; the lower is the end user data rate but also the lower the required investments.

Free space optics (FSO) is the technique of point-to-point laser communications. FSO is considered as a high-bandwidth wireless alternative to fiber optic cabling for short-haul access distances of 4 km or less. The primary advantages of FSO over fiber are its rapid deployment time and significant cost savings. On top of that, lasers offer the advantage of not requiring licensing frequency spectrum anywhere in the world. As a consequence of the visual communication principle involved, lasers require a direct line-of-sight (LOS) between transmitter and receiver. Because of the small lens size and tightly focused beam, free space optics is sensitive to vibrations, and the signal can easily be momentarily blocked by birds. Also fog can dramatically degrade the quality of the connection [15].

Satellites are successfully used for intercontinental telecommunications as well as television broadcasting. By the end of 2000, more than 600 satellites were deployed [16]. Space-related industries grow continuously and satellite as well as earth segment construction represent multi-billion dollar businesses. Also, bidirectional broadband fixed wireless access via a geostationary earth orbit (GEO) satellite is commercially offered. However, due to the long distance, GEO satellites suffer from round trip delays exceeding 0.25 seconds. Moreover, a high upstream bandwidth can only be achieved with large dish antennas. For example, the Astra Return Channel System (ARCS) uses dishes with a diameter of up to 1.2 m, for maximum uplink data rates of 2 Mbps [17].

To overcome the shortcomings of GEO satellite systems, several competing low earth orbit (LEO) satellite systems, at an altitude ranging between 500 km and 2000 km, were under construction in the late 90s. Examples include Teledesic [18], Iridium [19], Globalstar [20], and Skybridge [21]. These systems could accommodate shorter round trip delays and increased upstream data rate, at the cost of a large satellite constellation and a satellite dish that follows the moving satellites. None of these developments have lead to commercially viable solution.

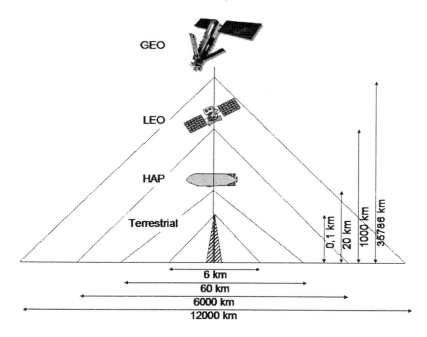

Figure 1.7. Height versus coverage area.

As an alternative approach to GEO satellites, high altitude platforms (HAPs) have also been proposed [22]. A HAP consists of a base station that is put on a specially designed aircraft, or a balloon, that is kept permanently above a coverage area. HAPs are typically located in the stratosphere, between 16 km and 22 km above the ground, and serve an area of approximately 50 km in diameter. Compared to satellite systems, they feature a reduced round-trip time, a higher data rate, a limited infrastructure investment, and a simplified end user terminal. HAPs need far less base stations than terrestrial solutions, due to their capability to serve coverage area. However, the complexity of a single base station is a lot higher. The trade-off between the height of a base station and its coverage area is illustrated in *Figure 1.7*. Due to the high frequencies used for broadband access over satellites and high altitude platforms, they all suffer from degrading communication due to rain fading.

Because power lines are omnipresent, power line communication (PLC) has been proposed as an attractive broadband access alternative. PLC has the potential of offering data rates between 10 and 100 Mbps, by using bandwidths up to 30 MHz [23]. Because of these high bandwidths, power line communication can not propagate through the medium voltage to low voltage transformers. As a consequence, a typical power line communication network has the architecture of *Figure 1.8*. The communication signal is injected on the power line after the transformer and traverses the last hundreds of meters to the home and inside the home to the end user equipment. Because of its bus architecture, the communication capacity on the power line has to be shared between the up to a few hundreds of homes connected to a single transformer.

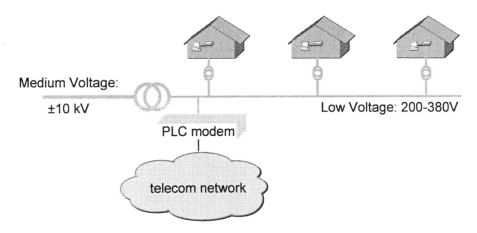

Figure 1.8. Power line communication network architecture.

Although power line communication has been successfully demonstrated and commercial products are available, doubts still exist about its viability as a wide scale alternative. For example, the power line plant can be considered as a large antenna, resulting in uncertainties with respect to the sensitivity to interference as well as about unwanted emissions. Furthermore, the conditioning of the power plant could raise similar challenges as encountered with the transformation of the cable network [24].

By reviewing these alternatives, it becomes clear that broadband fixed wireless access has the best potential to become the third broadband access technology. In developed countries, where existing wired networks are available, it will be used where bidirectional cable networks are not present, and digital subscriber lines can not be economically upgraded to broadband performance. In developing countries, it can be expected that no wired network will be deployed and that broadband fixed wireless access will become the mainstream access technology.

A large variety of broadband fixed wireless access technologies exist. The major distinction can be made based on their operating frequency, which can be either above or below 11 GHz. In the former case, a cellular point-to-multipoint line of-sight (LOS) network is established. An example of such a system is the Local Multipoint Distribution Service (LMDS), operating in frequencies between 27.5 GHz and 31.3 GHz [25]. By using highly directive antennas, the effective range of LMDS is about 3 miles (5 km), with data rates of up to 50 Mbps downstream and 10 Mbps upstream. Because of the high frequencies and narrow-beam antennas, the equipment is relatively costly. Moreover, the system will also suffer from rain fading. Sub-11 GHz BFWA, on the other hand, offers less available bandwidth, but features non-line-of-sight (NLOS) operation and cheaper terminals. Therefore, it is expected that sub-11 GHz operation will be dominant for residential and small office applications.

1.4 THE BFWA MARKET

A large number of market research studies have been performed by various companies. Although they differ in the detailed figures, they generally share the same major conclusions:

1. Broadband fixed wireless access, operating below 11 GHz, will become the third access technology with a global market share exceeding 5%.
2. The annual global market value for BFWA will surpass $1 billion from 2006 onwards.

According to *Table 1.1*, which represents a consolidation of the available figures in 2003, a major ramp up of the market may be expected in 2004, when standard based products become available. In more recent predictions, this moment is shifted to 2005, or even later, because of the delay in the development of standard-compliant products. To realize this market take-up, analysts indicate that several conditions need to be fulfilled.

Table 1.1. The BFWA market.

Worldwide BFWA Sub 11 Ghz Market	2001	2002	2003	2004	2005	2006	2007
Cumulative Worldwide Volume (m$)	720	905	1076	2100	3300	5000	11000
Volume (m$) / Year	229	185	171	1024	1200	1700	6000
Cumulative Number of lines (Klines)	876	1140	1482	4042	7471	13138	34566
Volume (Klines) / Year	279	264	342	2560	3429	5667	21429

Standardization of the technology should eliminate the confusion in the market caused by the many proprietary solutions. It would also enable interoperability between the equipment of different manufacturers, and, as such, stimulate competition.

This increased competition should lead to a considerable price drop of the customer premises equipment. Short term target prices below $350 are currently quoted.

The first generation of proprietary broadband fixed wireless access equipment featured a high operational cost. Especially the many truck rolls, needed to install and maintain the equipment, killed several business plans. The poor support of NLOS communication was the major underlying factor for these truck rolls.

However, technology is not sufficient; also the spectrum must be available. Therefore, regulators should take actions to free up BFWA spectrum. As several licenses on these spectra are already awarded but not yet used, active steps of the regulators might be required in several markets to recall the spectrum.

As the deployment of BFWA requires an upfront investment, the technology will only thrive in an improved economical climate. Also, competition amongst operators could accelerate the take up of the technology.

Concluding, it is expected that BFWA operating below 11 GHz will become the third access technology. The major critical success factor for this market take-up is the standardization that was recently completed.

1.5 OUTLINE OF THE BOOK

The recent standardization of sub-11GHz BFWA technology has spurred a renewed interest in broadband fixed wireless access, not only from operators and equipment manufacturers but also in academic research institutes. As a consequence, an increasing number of communication engineers get in touch with this new technology. We felt that a good overview of the state-of-the-art BFWA technology and its system implications was not available. This book aims to fill this gap. It contains nine chapters, the first of which is this introductory chapter.

Chapter 2 reviews the various standards and technologies that have been proposed for BFWA. In particular, the IEEE 802.11 standard, which was originally developed for indoor WLANs, has recently emerged as a wireless access technology for (local-area) hot spots. Furthermore, besides an FDD/WCDMA mode, which is intended for mobile users, the IMT-2000 standard also contains a TDD/TD-CDMA mode, which is intended for broadband access for stationary users. Finally, the IEEE is currently in the process of creating a new standard for high-rate mobile access within the IEEE 802.20 Mobile Broadband Wireless Access (MBWA) Working Group. Nevertheless, we come to the conclusion that the primary standards for BFWA have been realized by the IEEE 802.16 Broadband Wireless Access Working Group, which will be the focus of the remainder of the book.

Chapter 3 studies the BFWA radio channel, which constitutes a major challenge for the design of a BFWA radio system. A fundamental understanding of the main channel impairments is essential to come up with an optimal design of any radio system. This chapter gradually introduces the different impairments of the BFWA radio channel, including path loss, time-invariant multipath propagation, temporal variations, and angle dispersion.

Quality-of-Service (QoS) is an essential aspect of a BFWA system. The IEEE 802.16 BFWA data link layer, which includes the Medium Access Control (MAC) protocol that plays a key role in guaranteeing QoS, is described in Chapter 4. Furthermore, a performance assessment of the BFWA data link layer is included. Finally, service specific convergence layers are shown to constitute the interface between services and the data link layer.

The data link layer of IEEE 802.16 sits on top of multiple physical layers, each of them satisfying different needs in terms of the operating environment. Chapter 5 provides an overview of these different physical layers, emphasizing the employed modulation and coding schemes. One of these layers is suited for line-of-sight communications between 10 and 60 GHz, whereas the other three enable non-line-of-sight communications between 2 and 11 GHz. Due to its dominance on the market, the main focus will be on the OFDM physical layer. Finally, the main OFDM physical layer extensions to support mobile next to fixed subscriber stations are pointed out.

Chapter 6 presents the trade-offs for optimized implementations of IEEE 802.16 compliant terminals. First we elaborate on the overall architecture of the terminal, with a special focus on the interfaces between the hardware and the software and between the digital and analog processing. Next, we discuss the baseband receiver implementation. Besides the frequency domain equalization structure, we also pay attention to channel estimation and timing and frequency synchronization techniques. Finally we investigate the approach of a low-cost direct conversion radio front-end and the compensation of the resulting non-idealities in the digital domain. Combining these implementation techniques with advanced integration technologies will enable low cost BFWA equipment end will even lead to portable solutions.

In order to meet the data rate and QoS requirements of truly broadband wireless services, BFWA was one of the first application fields where the use of smart multi-antenna communication techniques was heavily promoted. Also the IEEE 802.16 standard includes the necessary features to support these smart antenna systems, both at the base stations and subscriber stations. After reviewing their potential gains and presenting a taxonomy of such smart antenna techniques, Chapter 7 points out their application in the IEEE 802.16 standard, emphasizing MIMO-OFDM processing techniques.

Chapter 8 highlights three technologies that could further reduce the total cost of ownership for the operator of a BFWA network. The first technology, autodirecting antennas, will simplify the installation of a BFWA terminal and reduce the number of truck rolls for the operator. Second, the bridging between BFWA and WLAN will enable wireless home gateways and hot-spots with wireless feeds. Finally, pay-as-you-grow infrastructure combines multi-hop communication and advanced antenna concepts to minimize the required capital for deploying and evolving a BFWA network. We believe that the above three technologies are crucial for reducing the total cost of ownership reduction and hence for the adoption of the BFWA technology.

Finally, Chapter 9 draws the main conclusions of this book.

1.6 REFERENCES

[1] G.C. Corazza, "Marconi's History", Proceedings of the IEEE, Vol. 86, No. 7, pp. 1307-1311, July 1998
[2] M. Kibati, "Wireless Local Loop in developing countries: Is it too soon for data? The case of Kenya", Master thesis, Master of Science in Technology and Policy, Massachusetts Institute of Technology, May, 1999.

[3] T.P. McGarty, "The Economic Viability of Wireless Local Loop and its Impact on Universal Service", Columbia University CITI seminar on "The Role of Wireless Communications in Delivering Universal Service", October 30, 1996.

[4] I. Dorros, "ISDN", IEEE Communications Magazine, Vol. 19, March 1981, pp. 16-19.

[5] http://www.clickz.com/stats/sectors/geographics/

[6] D.L. Waring, J.W. Lechleider, T.R. Hsing, "Digital Subscriber Line Technology Facilitates a Graceful Transition from Copper to Fiber", IEEE Communications Magazine, Vol. 29, March 1991, pp. 96-103.

[7] D. Fellows, D. Jones, "DOCSIS cable modem technology", IEEE Communications Magazine, Vol. 39, issue 3, March 2001, pp. 202-209.

[8] D.G. Mestdagh, M. R. Isaksson, P. Odling, "Zipper VDSL: a solution for robust duplex communication over telephone lines", IEEE Communications Magazine, Volume: 38, Issue: 5, pp. 90-96, May 2000.

[9] www.oecd.org

[10] P. Reusens, D. Van Bruyssel, J. Sevenhans, S. Van Den Bergh, B. Van Nimmen, P. Spruyt, "A Practical ADSL Technology Following a Decade of Effort", IEEE Communications Magazine, pp. 145-151, October 2001.

[11] T.R. Rowbotham, "Local loop developments in the UK", IEEE Communications Magazine, Vol.29, Issue 3, pp. 50-59, March 1991.

[12] S. Dravida, D. Gupta, S. Nanda, K. Rege, J. Strombosky, M. Tandon, "Broadband Access over Cable for Next-Generation Services: A Distributed Switch Architecture", IEEE Communications Magazine, pp. 116-124, August 2002.

[13] FS-VDSL Specification, Part 1: Operator Requirements, http://www.fs-vdsl.net/Specifications/FS-VDSL_PART1_V1-0_Final.pdf, June 5, 2002.

[14] EURESCOM, Project P614 Implementation Strategies for Advanced Access Networks, Deliverable 8: Elaboration of common FTTH guidelines, http://www.eurescom.de/~pub-deliverables/P600-series/P614/d8/d8.pdf, April 1999.

[15] I.I. Kim and E. Korevaar, "Availability of Free Space Optics (FSO) and hybrid FSO/RF systems, SPIE ITCOM 2001/ Wireless Communications IV, August 21, 2001.

[16] R. Struzak, "Basics of satellite-based and high-elevation platform-based radio links" School on Digital Radio Communications for Research and Training in Developing Countries, Trieste, Italy, 9 - 27 Feb. 2004.

[17] J. Neale, R. Green, J. Landovskis, "Interactive Channel for Multimedia Satellite Networks", IEEE Communications Magazine, Vol. 39, Nr. 3, pp. 192-198, March 2001

[18] M.A. Sturza, "Architecture of the TELEDESIC Satellite System", Proceedings of the International Mobile Satellite Conference, 1995.

[19] S.R. Pratt, R.A. Raines, C.E. Fossa Jr, M.A. Temple, "An Operational and Performance Overview of the IRIDIUM Low Earth Orbit Satellite System", IEEE Communications Surveys, http://www.comsoc.org/pubs/surveys, pp. 2-10, Second Quarter 1999.

[20] R.A. Wiedeman, A.J.Viterbi, "The Globalstar mobile satellite system for worldwide personal communications", JPL, Proceedings of the Third International Mobile Satellite Conference (IMSC 1993) p 291-296.

[21] D.J. Bem, T.W. Wieckowski, R.J. Zielinski, "Broadband Satellite Systems", IEEE Communications Surveys & Tutorials, vol. 3 no. 1, pp. 2-15, http://www.comsoc.org/pubs/surveys, First Quarter 2000

[22] T.C. Tozer and D. Grace, "High-altitude platforms for wireless communications", Electronics & Communication Engineering Journal, June 2001, pp. 127-137.

[23] R. Tongia, "Promises and False Promises of PowerLine Carrier (PLC) Broadband Communications – A Techno-Economic Analysis", The 31st Research Conference on Communication, Information and Internet Policy, September 19-21, 2003

[24] J. Löcher, G. Varjú, "Survey of the present-day state of the power line telecommunication, PLT technology", Postgraduate Conference on Electric Power Systems, Budapest, August 12-13. 2002.

[25] H. Sari, Broadband radio access to homes and businesses, IEEE Computer Networks Vol. 31 (1999) page 379-393.

Chapter 2

A Plethora of BFWA Standards
BFWA Standard Technologies

Frederik Petré

2.1 INTRODUCTION

This chapter takes a closer look at the available standards for Broadband Fixed Wireless Access (BFWA). As pointed out in Chapter 1, standardization is critical to align the market, and to allow the development of interoperable equipment. In contrast to popular belief, we are convinced that standardization does not inhibit creativity but rather enables a flourishing market, in which innovative products can thrive. As standardization is an important market creation tool, it should also be pointed out that market forums, like WiMax and WiFi, play an essential role in promoting and certifying interoperable products based on standards.

Several standards are used by BFWA equipment manufacturers, although not all were originally developed with BFWA as their primary target application. The IEEE 802.11 standard was originally developed for indoor Wireless Local Area Networks (WLANs), although also frequently used for point-to-point bridging of wireline local area networks. Recently, it also has emerged as a wireless access technology for (local area) hot spots. The European Telecommunication Standardization Institute (ETSI) has defined an alternative standard, namely High-Performance Local Area Networks (HIPERLAN). However, it did not receive market acceptance, and, hence, will not be considered further. The IMT-2000 standard contains both a Frequency-Division Duplex (FDD) and a Time-Division Duplex (TDD) mode. Where the former is based on pure Wideband Code-Division Multiple Access (WCDMA), the latter includes an additional Time-Division Multiple Access (TDMA) component, called TD-CDMA. Although the main focus of

IMT-2000 is on the FDD/WCDMA mode for moving users, there is a growing interest in the TDD/TD-CDMA mode, which is intended for broadband access for stationary users. Recently, the IEEE started a standardization effort for high-rate mobile access in the 802.20 Working Group. Nevertheless, the primary standards for BFWA have been realized by the IEEE 802.16 Broadband Wireless Access Working Group. The line-of-sight (LOS) systems in the frequency bands above 11 GHz have been defined in the baseline IEEE 802.16 standard. An alternative system has been defined in the ETSI High-Performance Access (HIPERACCESS) standard. Non-line-of-sight (NLOS) systems below 11 GHz follow the IEEE 802.16a standard. Due to a close collaboration between ETSI and IEEE, the High-Performance Metropolitan Area Network (HIPERMAN) standard is a subset of the IEEE 802.16a standard. In 2004, a consolidation of all 802.16 standards was realized, which resulted in the 802.16-2004 revision.

In the following sections, we will discuss each of these standards in more detail. Section 2.2 discusses the IEEE 802.11 WLAN standard, whereas Section 2.3 briefly reviews the IMT-2000 TDD/TD-CDMA standard. Section 2.4 introduces the IEEE 802.20 Mobile Broadband Wireless Access (MBWA) standard. Sections 2.5 and 2.6 briefly describe the true BFWA standards, with Section 2.5 focusing on the 10-66 GHz IEEE 802.16/HIPERACCESS standard, and with Section 2.6 focusing on the sub-11 GHz IEEE 802.16a/HIPERMAN standard. Finally, Section 2.7 comments on the convergence of different wireless access technologies towards a common wireless platform.

2.2 THE IEEE 802.11 WIRELESS LAN STANDARD

The IEEE 802.11 WLAN standard was originally developed for the 2.4 GHz unlicensed spectrum, and later on extended to the frequency bands between 5 and 6 GHz [1][2]. As illustrated in *Figure 2.1*, the standard supports both infrastructure-based networks as well as ad-hoc networking, with a maximum data rate of 54 Mbps per cell. The communication range is typically limited to 50 m for indoor use and to 300 m for outdoor applications. The price of 802.11-compliant equipment is eroding very fast, and the installation can easily be performed by an end user. Up till recently, no mechanism was available in the standard to provide Quality-of-Service (QoS) guarantees. A recently finalized effort in the 802.11e Task Group has tried to resolve this shortcoming. Moreover, Siemens has recently launched a product that offers proprietary QoS features for industrial applications [3]. Security is a recent enhancement of the 802.11 standard. Moreover, the Working Group is looking at extensions towards multi-hop or mesh networking, and towards higher speeds, targeting more than 100 Mbps effective data rate. Multi-antenna techniques will be most likely an essential part of this high data rate extension. Several companies have already demonstrated the capabilities of multi-antenna techniques for increasing both the capacity and the range of a network.

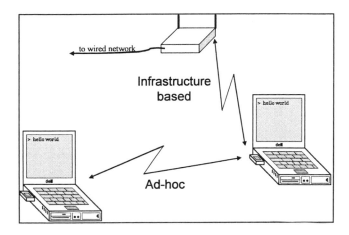

Figure 2.1. Infrastructure-based versus ad-hoc networking scenario.

The extensions of the base standard, which originally provided 1 and 2 Mbps in the 2.4 GHz band, are indicated with a letter. This resulted in the following confusing "letter soup":

- 802.11a is the extension of the standard towards 54 Mbps communications in the frequency bands between 5 and 6 GHz.
- 802.11b extends the base standard with a 5.5 Mbps and an 11 Mbps mode.
- 802.11c defines the bridging operation with other local area networks.
- 802.11d added features to the hopping pattern and the management information base, to support the use of the standard outside the United States (US).
- 802.11e is a currently finalized Task Group that has been working on the definition of mechanisms to guarantee QoS.
- 802.11f defines recommended practices for implementing a protocol between 802.11 access points. As such, it is guaranteed that access points of multiple vendors will interoperate in a WLAN.
- The 802.11g Task Group extended the data rate of WLANs in the 2.4 GHz band towards 54 Mbps.

- 802.11h defined the dynamic frequency selection for the 5 to 6 GHz bands. This was crucial for Europe, where WLANs are only the secondary service in this frequency band.
- The 802.11i standard provided enhanced security mechanisms. These have become a major requirement after it was demontrated that the initial Wired Equivalent Privacy (WEP) algorithm could be broken.
- The 802.11j group provided channel selection for the 4.9 GHz band, which is available in Japan.
- 802.11k extends the radio resource measurements of an entity, and provides the mechanisms to communicate those measurements to other entities and higher layers.
- 802.11m is an active Task Group to maintain and correct the original baseline standard.
- The 802.11n Task Group aims at defining an additional physical layer (PHY) by the end of 2006, to increase the effective data rate to at least 100 Mbps.
- 802.11r is working towards provisioning fast roaming between access points.
- 802.11s has started mid 2004, to study mesh networking.
- 802.11t is discussing the definition of standardized methods to evaluate the wireless performance of an 802.11 system.

As pointed out in *Table 2.1*, four of these standard extensions define additional physical communication modes with increasing data rates and bandwidth efficiencies. The base standard defined 1 and 2 Mbps modes based on Direct-Sequence Spread-Spectrum (DSSS). The 802.11b enhancement added 5.5 and 11 Mbps modes, which use Complementary Code Keying (CCK). The 802.11a extension defines an Orthogonal Frequency Division Multiplexing (OFDM) mode in the 5 GHz band that provides up to 54 Mbps. In the 802.11g standard, this OFDM mode was also made available for the 2.4 GHz band. Finally, the 802.11n standard is still under construction and aims at increasing the effective data rate, measured at the MAC Service Access Point (SAP), i.e. on top of the MAC layer, above 100 Mbps. This minimum throughput requirement represents a four times leap in throughput performance compared to existing 802.11a/g WLANs, as the 54 Mbps mode only results in an effective data rate of 25 Mbps. Since May 2005, there are only two surviving candidate proposals, namely WWiSE [4] and TgnSync [5], both of which share the idea of using Multi-Input Multi-Output (MIMO) OFDM communications, through the deployment of multiple antennas at both the transmitting and receiving side. Although both proposals rely on MIMO technology to reach the required high throughputs, they resort to different MIMO processing techniques. On the one hand, the WWiSE proposal follows a cautious and evolutionary approach, based on simple Alamouti Space-Time Block Coding. On the other hand, TgnSync supports more audacious and future-proof spatial multiplexing solutions, including joint transmit and receive processing techniques. The final IEEE 802.11n draft standard is expected to be finalized by September 2006.

Table 2.1. The IEEE 802.11 standard physical communication modes.

Standard	Frequency band (GHz)	Data rate (Mbps)	BW efficiency (bits/s/Hz)
802.11	2.4	1-2	0.05-0.1 (DS-SS)
802.11a	5	6-54	0.3-2.7
802.11b	2.4	5.5-11	0.275-0.55
802.11g	2.4	1-54	0.05-2.7
802.11n	2.4 & 5	> 100 (effective)	> 3

2.3 THE IMT-2000 TDD/TD-CDMA STANDARD

The IMT-2000 standard for a third generation (3G) mobile radio system comprises both an FDD and a TDD mode. The FDD mode uses a minimum spectrum of 5 MHz in each communication direction in the paired frequency bands between 1920 – 1980 MHz and between 2110 – 2170 MHz. The TDD mode uses a minimum spectrum of 5 MHz for up- and downlink communication in the unpaired frequency band between 1885 – 1920 MHz. Hence, two times 60 MHz of licensed spectrum is available for FDD operation, while 35 MHz of licensed spectrum is available for TDD operation.

The FDD mode, which is based on pure WCDMA at a chip rate of 3.84 Mcps, is mainly intended for mobile users in public macro and micro cell environments, supporting data rates up to 384 Kbps [6]. The TDD mode, which is based on TD-CDMA (a combination of TDMA and CDMA) at the same chip rate, is mainly intended for stationary users in public micro and pico cell environments, and for broadband fixed wireless access applications, supporting data rates up to 2 Mbps [7]. Although not practically supported by the system, a maximum user data rate of 3.3 Mbps can be achieved, in case no spreading is performed.

The system is based on a point-to-multipoint network architecture, and supports communication with variable QoS requirements. Furthermore, the system entails several PHY procedures that are essential for system operation, for instance, to support seamless handover and roaming, such as handover measurements, initial cell search, and random access. Based on a classical rake receiver, which is depicted in *Figure 2.2*, combined with powerful Forward Error Correction (FEC) coding, the system can handle multipath propagation and multiuser interference, only to some extent. In this case, the capacity that can be achieved with a reasonable bit error rate (BER) performance will be very limited, with typical values up to 20 % of the maximum system load. Based on advanced multiuser receivers at the base station for uplink communication, and based on advanced chip equalizer receivers at the mobile station for downlink communication, the system can effectively handle intra-cell interference (interference between users residing in the same cell) as well as non-line-of-sight multipath communications [9]. In this case, the system capacity will be limited by the inter-cell interference, that is the interference originating from users residing in neighbouring cells.

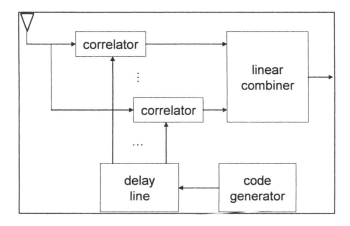

Figure 2.2. Classical rake receiver collects signal energy in different propagation paths.

Implementations of the TDD/TD-CDMA mode are available on the market, for instance from IP Wireless [8]. Because they follow the principle of a direct coverage from the base station towards the computer, even indoor, they can be easily installed by the end user. However, the main shortcoming of this technology is the relatively low data rate. Consequently, this is one of the major areas for future improvements. More specifically, the High-Speed Downlink Packet Access (HSDPA) standard evolution, which is currently under investigation, aims to significantly reduce the end-to-end delay as well as to increase the effective data rate up to 10 Mbps for downlink packet data, by resorting to higher modulation formats and code rates, combined with a low spreading factor, and multiple parallel codes per user.

2.4 THE IEEE 802.20 MBWA STANDARD

In December 2002, the IEEE 802.20 Mobile Broadband Wireless Access (MBWA) Working Group was established with the mission of specifying a broadband packet-based air interface for mobile users with speeds up to 250 km/h [10]. The 802.20 MBWA system should operate in licensed bands below 3.5 GHz, supporting both FDD and TDD operation. Furthermore, it should support deployment in at least one of the following frequency arrangements: 1.25 MHz, 5 MHz, 10 MHz, 15 MHz, or 20 MHz in each direction for FDD operation, and 2.5 MHz, 5 MHz, 10 MHz, 20 MHz, 30 MHz, or 40 MHz for TDD operation. Targeting a sustained spectral efficiency well beyond 1 bps/Hz/cell, the peak user data rate should exceed 1 Mbps in the downlink and 300 Kbps in the uplink. Furthermore, the peak aggregate data rate per cell should exceed 4 Mbps in the downlink and 800 Kbps in the uplink. It should be emphasized that these targets are for a 1.25 MHz channel bandwidth, which corresponds to two 1.25 MHz paired frequency bands for FDD or a 2.5 MHz unpaired frequency band for TDD. For other channel bandwidths, the data rates should be scaled in proportion to the channel bandwidth.

Although the Working Group has experienced a difficult start, it has finalized a system requirements document in July 2004 [11]. Currently, the Working Group is coming to an agreement on channel models for system simulations [12] and evaluation criteria [13]. Furthermore, it is finalizing the discussions on the technology selection process, such that a call for proposals can be issued in 2006. Although Flarion with their Flash OFDM system [14] were the main promoters of this Working Group, all options for the Medium Access Control (MAC) layer and the PHY of the system are still open. Block-spread CDMA transmission techniques, which judiciously combine CDMA and OFDM concepts, seem very appealing for the IEEE 802.20 MBWA system, since they inherit the attractive features of both worlds [9]. They can be divided into two main families.

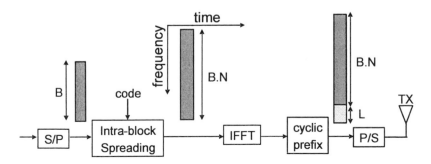

Figure 2.3. Intra-block-spread CDMA techniques spread information along frequency axis.

On the one hand, so-called intra-block-spread CDMA techniques that perform spreading of the information blocks along the frequency axis, such as Multi-Carrier CDMA (MC–CDMA) and Single-Carrier CDMA (SC-CDMA) are well suited for high mobility users in macro and micro cell environments. A general transmitter block diagram of these intra-block-spread CDMA techniques is shown in *Figure 2.3*.

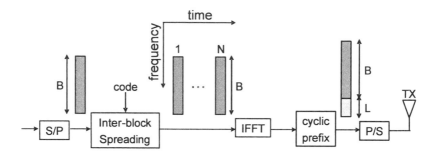

Figure 2.4. Inter-block-spread CDMA techniques spread information along time axis.

On the other hand, so-called inter-block-spread CDMA techniques that perform spreading of the information blocks along the time axis, such as Multi-Carrier Block-Spread CDMA (MCBS–CDMA) and Single-Carrier Block-Spread CDMA (SCBS–CDMA), are well suited for stationary to low mobility users in micro and pico cell environments. A general transmitter block diagram of these inter-block-spread CDMA techniques is shown in *Figure 2.4*.

Finally, multiple antenna techniques, both at the base and mobile station, are key to meet the data rate and QoS requirements.

2.5 THE IEEE 802.16/HIPERACCESS STANDARD

The dominant BFWA standard is the IEEE 802.16 standard, as it was specifically designed for this application. The base standard defines a point-to-multipoint network that operates in licensed frequency bands between 10 and 66 GHz [15] [3]. In Europe, ETSI specified HIPERACCESS, a similar yet different standard for the same application.

The high operating frequencies make that only line-of-sight operation with highly directive antennas is feasible. Because narrow-beam antennas need very careful installation, this normally requires the involvement of the operator. The high frequencies also increase the sensitivity of the system to weather conditions, or, in particular, to rain fading. They further increase the cost of the terminal because of the difficulty of integrating such high-frequency circuits.

The 802.16 and HIPERACCESS standards define communication in channel bandwidths of 25 to 28 MHz, and data rates exceeding 120 Mbps. The MAC layer is inspired on the cable modem standard DOCSIS, and, hence, has extensive features for managing the QoS of connections.

With its high data rates, but also with its high costs (with typical values between 500 and 15000 Euros), the 802.16 system is typically suited for business applications and backhaul of wireless networks.

To improve the reliability of the network, mesh networks have been, until now unsuccessfully, promoted [17]. For example, Radiant networks (sold to LamTech) developed a roof-mounted user terminal with four vertically stacked highly directional antennas that can be steered with full freedom of +/-360° rotation in the horizontal plane.

2.6 THE IEEE 802.16A/HIPERMAN STANDARD

The 802.16a standard inherits the MAC layer from the base standard, and adds PHY communication schemes that are suited for non-line-of-sight communication in frequency bands below 11 GHz [3]. Both licensed (mainly 2.5-2.69 GHz in the US and 3.4-3.6 GHz in Europe and the rest of the world) and unlicensed spectrum (5.725-5.850 GHz as the most important band) is supported. The ETSI HIPERMAN committee closely cooperated with the IEEE Working Group, and defined the OFDM subset of the 802.16a standard as a European standard [4].

The relatively low operating frequencies bring several advantages to the IEEE 802.16a system: non-line-of-sight operation, low terminal cost (with typical values less than 400 Euros), and ease of installation. Consequently, 802.16a is ideally suited for broadband fixed wireless access to small and medium enterprises, to small offices and home offices, and to residential customers. It can also be used for backhaul of 802.11 hot spots. The standard supports multiple bandwidths and data rates up to 104 Mbps in a 28 MHz channel bandwidth. However, a typical deployment would rather use a 3.5 MHz channel bandwidth, resulting in approximately 13 Mbps, or a 5 MHz channel bandwidth, resulting in just over 18 Mbps. Recent performance benchmarks have shown, however, that the total average downlink throughput can be expected to be between 3 and 7 Mbps over a 5 MHz bandwidth, with the lower rates corresponding to having a single receive antenna and three-sector cells, and the higher rates corresponding to having two receive antennas and six-sector cells [5].

Besides the centrally scheduled point-to-multipoint architecture of 802.16, the standard also adds a mesh mode. This mode allows extending the reach of the network, at the cost of the network capacity. A lot of research is still ongoing to optimize this range versus capacity trade off for mesh networks. Also guaranteed QoS in a mesh network is an active research topic. In addition, multiple antenna techniques, which are another means of improving range and capacity, are actively studied.

In June 2004, the IEEE 802.16 standard and the IEEE 802.16a standard were consolidated in a unified IEEE 802.16d standard, which was published as the IEEE 802.16-2004 in December, 2004 [1]. Currently, the IEEE 802.16e Task Group is defining extensions to this standard for moving terminals with mobile speeds up to 120 km/h [14].

2.7 TOWARDS A COMMON WIRELESS PLATFORM

The previous sections showed that multiple technologies compete to provide broadband wireless access. We do not believe that a single technology will emerge as a sole winner. Indeed, the technologies are all optimized for a particular operating environment, and are suboptimal in order situations. *Figure 2.5* illustrates this by providing an overview of current and future wireless data communication standards as a function of maximum data rate and typical cell range.

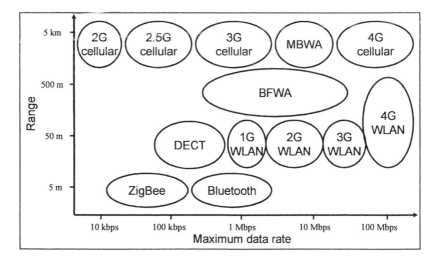

Figure 2.5. Different wireless data communication standards complement each other for maximum data rate versus cell range.

Hence, future-generation wireless systems will integrate different existing and new evolving wireless access systems on a common IP-based platform, to complement each other for different service requirements and radio environments [22]. To enable seamless and transparent interworking between these different wireless access modes, multistandard functionality is needed, especially at the terminal side. As such the different technologies can be combined to realize ubiquitous connectivity (see *Figure 2.6*): IMT-2000 or IEEE 802.20 when travelling at high speed, 802.11 at the office and a combination of 802.11 and 802.16 at home or within reach of a public hot spot.

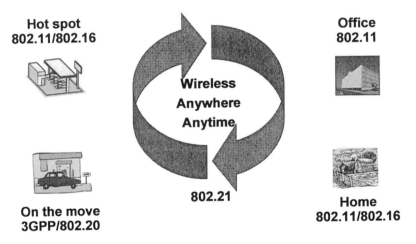

Hot spot
802.11/802.16

Office
802.11

Wireless
Anywhere
Anytime

On the move
3GPP/802.20

802.21

Home
802.11/802.16

Figure 2.6. Wireless anywhere and anytime.

Since IEEE shares this vision, it has established the IEEE 802.21 Media Independent Handoff Working Group to study the handoff between these various technologies [23].

2.8 REFERENCES

[1] R. Van Nee, G. Awater, M. Morikura, H. Takanashi, M. Webster, K. W. Halford, "New High-Rate Wireless LAN Standards", IEEE Communications Magazine, Vol. 37, No. 12, pp. 82-88, December 1999.

[2] A. Doufexi, S. Armour, M. Butler, A. Nix, D. Bull, J. McGeehan, P. Karlsson, "A Comparison of the HIPERLAN/2 and IEEE 802.11a Wireless LAN Standards", IEEE Communications Magazine, Vol. 40, No. 5, pp. 172-180, May 2002.

[3] Scalance W WLAN products, http://www.siemens.com

[4] C. Kose et al., "WWiSE proposal: High-Throughput Extension to the IEEE 802.11 standard", IEEE 802.11n Task Group, Doc. Nr. IEEE 802.11 – 05/0149r1, January 2005.

[5] S.A. Mujtaba et al., "TGnSync proposal: Technical Specification", IEEE 802.11n Task Group, Doc. Nr. IEEE 802.11 – 04/0889r44, March 2005.

[6] E. Dahlman, P. Beming, J. Knutsson, F. Ovesjö, M. Persson, C. Roobol, "WCDMA – The Radio Interface for Future Mobile Multimedia Communications", IEEE Transactions on Vehicular Technology, Vol. 47, No. 4, pp. 1105-1118, November 1998.

[7] M. Haardt, A. Klein, R. Koehn, S. Oestreich, M. Purat, V. Sommer, T. Ulrich, "The TD-CDMA Based UTRA TDD Mode", IEEE Journal on Selected Areas in Communications, Vol. 18, No. 8, pp. 1375-1385, August 2000.

[8] UMTS TD-CDMA solutions, http://www.ipwireless.com

[9] F. Petré, "Block-Spread CDMA for Broadband Cellular Networks", PhD Thesis, Katholieke Universiteit Leuven, Leuven, Belgium, December 2003.

[10] IEEE 802.20 Mobile Broadband Wireless Access Working Group, "IEEE 802.20 Five Criteria for Standards Development", December 2002.

[11] IEEE 802.20 Mobile Broadband Wireless Access Working Group, "System Requirements for IEEE 802.20 Mobile Broadband Wireless Access Systems", Version 14, July 2004.

[12] IEEE 802.20 Mobile Broadband Wireless Access Working Group, "Channel Models for IEEE 802.20 MBWA System Simulations", Version 9, Revision 1, July 2005.

[13] IEEE 802.20 Mobile Broadband Wireless Access Working Group, "IEEE 802.20 Evaluation Criteria", Version 17, August 2005.

[14] Flash-OFDM technology, http://www.flarion.com

[15] IEEE 802.16-2001, "IEEE Standard for Local and Metropolitan Area Networks – Part 16: Air Interface for Fixed Broadband Wireless Access Systems", April 2002.

[16] C. Eklund, R. B. Marks, K. L. Stanwood, S. Wang, "IEEE Standard 802.16: A Technical Overview of the WirelessMAN Air Interface for Broadband Wireless Access", IEEE Communications Magazine, Vol. 40, No. 6, pp. 98-107, June 2002.

[17] P. Whitehead, "Mesh Networks; a New Architecture for Broadband Wireless Access Systems", RAWCON 2000, pp 43-46, September 2000.

[18] IEEE 802.16-2004, "IEEE Standard for Local and Metropolitan Area Networks – Part 16: Air Interface for Fixed Broadband Wireless Access Systems", Revision of IEEE 802.16-2001, December 2004.

[19] I. Koffman, V. Roman, "Broadband Wireless Access Solutions Based on OFDM Access in IEEE 802.16", IEEE Communications Magazine, Vol. 40, No. 4, pp. 96-103, April 2002.

[20] A. Gosh, D. R. Wolter, J. G. Andrews, R. Chen, "Broadband Wireless Access with WiMax/802.16: Current Performance Benchmarks and Future Potential", IEEE Communications Magazine, Vol. 43, No. 2, pp. 129-136, February 2005.

[21] IEEE P802.16e/D8-2005, "Draft Amendment to IEEE Standard for Local and Metropolitan Area Networks – Part 16: Air Interface for Fixed Broadband Wireless Access Systems – Physical and Medium Access Control Layers for Combined Fixed and Mobile Operation in Licensed Bands", May 2005.

[22] F. Petré, A. Kondoz, S. Kaiser, A. Pandharipande, "Guest Editorial", Special Issue on Reconfigurable Radio for Future Generation Wireless Systems, EURASIP Journal on Wireless Communications and Networking (JWCN), Vol. 2005, No. 3, pp. 271-274, August 2005.

[23] IEEE 802.21 Media Independent Handoff Working Group, "IEEE 802.21 Five Criteria for Standards Development", March 2004.

Chapter 3

Unraveling the BFWA Propagation Environment
The BFWA Radio Channel

Frederik Petré

With contributions by Rafael Torres and Nadia Khaled

3.1 INTRODUCTION

A thorough understanding of its radio propagation channel is of paramount importance in the design of any wireless system. To enable reliable data communication over such a radio propagation channel, the transmitter and the receiver should be designed to cope with its impairments. Hence, a realistic model of the radio propagation channel is essential to design and evaluate different transmitter and receiver algorithms.

Two basic types of channel models exist that serve different purposes, namely deterministic versus stochastic models. On the one hand, a deterministic model relies on the actual propagation mechanisms to capture the physical propagation paths. Such a model generates an accurate channel response for a specific propagation environment, based on stored measurement results or ray tracing simulation results. Although well-suited for site-specific system deployments and network planning, it is less appropriate for system design and testing. On the other hand, a stochastic model generates channel responses as specific realisations of a multidimensional distribution, which statistically describes the set of all possible realisations of propagation channels. With such a stochastic model the performance of a complete communication system can be determined through site-independent Monte-Carlo simulations.

Since we focus on the BFWA application, we are especially interested in the BFWA radio channel, which is characterized by considerable multipath propagation [1] in combination with limited mobility in the propagation environment. Multipath propagation causes spreading of the signal along three different dimensions: delay, (Doppler) frequency, and angle [3],[4]. As we will discuss in more detail in the subsequent sections, the signal spreading along these different dimensions is characterized by the delay spread, the Doppler spread, and the angle spread, respectively. The delay spread describes the phenomenon where a single transmitted impulse in the time domain arrives at the receiver as several time-shifted and scaled versions of this impulse. The Doppler spread, which requires motion of the transmitter, the receiver, or the environment on top of multipath propagation, characterizes the situation where a pure frequency tone is spread over a finite spectral bandwidth. Finally, the angle spread captures the effect where a signal does not arrive at the receiver from one single direction, but rather from a multitude of directions.

In this chapter, we review the main impairments of the BFWA radio channel, and present stochastic channel models to capture these impairments in simulations. In order to study the propagation behaviour in BFWA radio channels, a number of realistic system assumptions have to be made [1]:

- the cell radius is typically between 1 and 10 km;
- the Base Transceiver System (BTS), or base station, antenna has a height between 15 and 40 meters;
- the Customer Premise Equipment (CPE) antenna has a height between 2 and 10 meters;
- the CPE antenna has a 3 dB beamwidth that is larger than 20°;
- non-line-of-sight propagation;
- and typical transmission bandwidths are 3.5 MHz.

This chapter is organized as follows. Section 3.2 discusses the path loss experienced within BFWA channels, including outdoor path loss as well as outdoor-to-indoor propagation loss. Section 3.3 quantifies the characteristics of time-invariant multipath propagation. Section 3.4 presents the temporal variation of a multipath channel and reports a measurement and statistical analysis campaign of the temporal variation of the BFWA channel at 3.5 GHz in suburban areas. Section 3.5 reveals the characteristics of multiple antenna BFWA channels, including angle dispersion both at the transmitter and the receiver. Finally, Section 3.6 summarizes the main conclusions.

3.2 PATH LOSS

Like any radio channel, the BFWA radio channel is characterized by two types of fading effects: large-scale and small-scale fading [3]. On the one hand, large-scale fading refers to the received signal power attenuation, or path loss, due to large changes in the distance between the transmitter and the receiver. Furthermore, the signal path between transmitter and receiver is often obstructed by objects with dimensions larger than a wavelength, which is referred to as shadowing. Hence, the path loss as a function of distance is generally described in terms of a mean path loss, caused by distance attenuation, and a log-normally distributed variation about the mean, caused by shadowing. On the other hand, small-scale fading refers to dramatic changes in the received signal power, due to small changes (as small as a half-wavelength) in the distance between transmitter and receiver. As will be explained in more detail in Section 3.3, small-scale fading manifests itself in the delay-dispersive and time-variant behaviour of the channel. In this section, we study the path loss behaviour of the BFWA channel, which comprises the outdoor path loss as well as the outdoor-to-indoor propagation loss.

This section is organized as follows. Subsection 3.2.1 describes an empirically based model for the outdoor path loss experienced in suburban environments. Subsection 3.2.2 discusses a propagation model to accurately predict outdoor-to-indoor propagation loss.

3.2.1 Outdoor path loss

Under the highly idealized free space propagation model, based on the following assumptions: (1) the region between the transmit and receive antennas is free of objects that may absorb or reflect RF energy, (2) the atmosphere behaves as a perfectly uniform and non-absorbing medium, and (3) the earth is infinitely far away from the propagating signal, the path loss (in dB) can be expressed as:

$$PL_s(d) = 20 \log\left(\frac{4\pi d}{\lambda}\right), \tag{3.1}$$

where d is the distance between the transmitter and the receiver, and λ is the wavelength of the propagating signal.

However, for most practical channels, like the BFWA radio channel, in which the region between the transmit and receive antennas is full of objects that may reflect, diffract, or scatter the RF energy, and, in which the signal propagation takes place in the atmosphere and near the ground, the free space propagation model is not adequate to fully describe the channel's path loss behaviour. Indeed, the path loss does not only depend on the distance and the RF frequency, but also on the type of propagation environment, e.g. the type of tree and building density, and the height of the BTS and CPE antennas. In general, the average path loss increases with tree and building density, while decreasing with the flatness of the propagation environment and the height of BTS and CPE antennas. For instance, Hong et al measured the path loss as a function of the distance between BTS and CPE for various CPE antenna heights between 6 m and 10 m above ground level [5]. The least-squares regressions fits to the measurement data clearly illustrate the trend of decreasing path loss exponent (slope of the fitted lines) for increasing CPE antenna heights.

The practicalities introduced above are taken into account by the path loss model proposed by the IEEE 802.16 standardization group, which is valid for propagation in suburban environments [1],[6]. The path loss (in dB) as a function of distance can be expressed as:

$$PL(d) = PL_s(d_0) + 10\gamma \log\left(\frac{d}{d_0}\right) + \Delta PL_f + \Delta PL_h + s \quad d \geq d_0 \quad (3.2)$$

where $PL_s(d_0)$ is the free space path loss at a reference distance $d_0 = 100m$, and, where γ, ΔPL_f, ΔPL_h, and s are characterized below. The path loss exponent γ, which depends on the terrain category and the BTS antenna height h_b (in meters), can be expressed as follows:

$$\gamma = a - bh_b + \frac{c}{h_b} \quad 10m \leq h_b \leq 80m \tag{3.3}$$

where a, b, and c, are empirically derived constants for different terrain categories. The numerical values of these model parameters are shown in Table 3.1 for three different terrain categories: worst, intermediate, and best terrain. The worst terrain category corresponds to hilly terrain with moderate-to-heavy tree densities, whereas the best terrain category corresponds to mostly flat terrain with light tree densities. The intermediate terrain category corresponds to hilly terrain with light tree densities or flat terrain with moderate-to-heavy tree densities.

Table 3.1. Numerical values of model parameters for three different terrain categories.

Model parameter	Worst terrain	Intermediate terrain	Best terrain
a	4.6	4	3.6
b (in m^{-1})	0.0075	0.0065	0.005
c (in m)	12.6	17.1	20

In order to use the model at other RF frequencies than 2 GHz, a frequency correction term ΔPL_f is introduced, which is given by:

$$\Delta PL_f = 6\log_{10}\left(\frac{f}{2000}\right) \tag{3.4}$$

where f is the carrier frequency in MHz. In order to use the model at other CPE antenna heights besides 2 m, a CPE antenna height correction term is introduced, which is given by:

$$\Delta PL_h = -10.8\log_{10}\left(\frac{h}{2}\right) \tag{3.5}$$

for the worst and intermediate terrain types, and by:

$$\Delta PL_h = -20\log_{10}\left(\frac{h}{2}\right) \tag{3.6}$$

for the best terrain type, where h is the CPE antenna height between 2 m and 10 m. Finally, the shadowing effect is represented by s, which follows a zero-mean Gaussian distribution. Typical values for the standard deviation of s are between 8.2 and 10.6 dB, depending on the terrain type and tree density.

3.2.2 Outdoor-to-indoor propagation

For our discussion in Subsection 3.2.1, we have implicitly assumed that the CPE is placed externally, such that outdoor propagation is the main propagation process. However, for CPEs in the form of, for instance, a PC card, outdoor-to-indoor propagation should be considered, which comprises three distinct propagation processes. Next to the outdoor propagation loss PL_{out}, there is also the building penetration loss PL_{pn} and the indoor propagation loss PL_{in}. The total path loss can then be expressed as the sum of the losses for these three individual parts:

$$PL_{tot} = PL_{out} + PL_{pn} + PL_{in}, \tag{3.7}$$

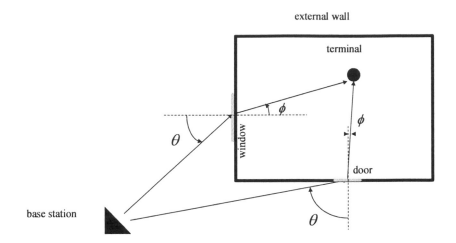

Figure 3.1. Outdoor-to-indoor propagation model according to Yuko Miura [7].

If there are wall openings on the building's external wall, such as doors and windows, it is reasonable to assume that the propagation paths through the wall openings dominate the outdoor-to-indoor propagation process, and, hence, determine the path loss at the CPE. According to Yuko Miura's outdoor-to-indoor propagation model [7], which is illustrated in *Figure 3.1*, the penetration loss for a particular propagation path k can be expressed as follows:

$$PL_{pn,k} = W_{e,k} + \Delta PL_{pn,out,k} + \Delta PL_{pn,in,k} \tag{3.8}$$

where $W_{e,k}$ is the loss across the wall opening with perpendicular penetration ($\theta_k = 0°$), $\Delta PL_{pn,out,k}$ is a first correction term to account for the outdoor angular dependency, and $\Delta PL_{pn,in,k}$ is a second correction term to account for the indoor angular dependency [7]. On the one hand, the outdoor angular dependency from BTS to wall opening can be expressed as follows:

$$\Delta PL_{pn,out,k} = WG_{e,k} \cdot (1 - \cos\theta_k)^2 \tag{3.9}$$

where θ_k is the grazing angle of the external wall, and $WG_{e,k}$ is the additional external loss in the wall opening when $\theta_k = 90°$. On the other hand, the indoor angular dependency from wall opening to CPE can be expressed as follows:

$$\Delta PL_{pn,in,k} = WG_{i,k} \cdot \sin\phi_k \tag{3.10}$$

where ϕ_k is the grazing angle of the internal wall, and $WG_{i,k}$ is the additional internal loss in the wall opening when $\phi_k=90°$. Typical values for $W_{e,k}$, $WG_{e,k}$, and $WG_{i,k}$ for a door and an RF frequency of 8.45 GHz are 17.2 dB, 20 dB, and 20 dB, respectively. If there are several wall openings, and, hence, several received propagation paths, the total penetration loss is the parallel combination of the penetration losses encountered by the individual propagation paths:

$$\frac{1}{PL_{pn}} = \sum_k \frac{1}{PL_{pn,k}} \tag{3.11}$$

Finally, following a logarithmic path loss model, the indoor propagation loss for a particular propagation path can be expressed as follows:

$$PL_{in}(d_{in}) = PL_s(d_{in,0}) + 10\alpha \log\left(\frac{d_{in}}{d_{in,0}}\right) + s_{in}, \tag{3. 12}$$

where d_{in} is the distance from the external wall to the CPE, $PL_s(d_{in,0})$ is the free space path loss at an indoor reference distance $d_{in,0} = 1m$, α is the indoor path loss exponent, which varies between 1.8 (for lightly obstructed environments with corridors) and 5 (for multi-floored buildings), and s_{in} is a zero-mean Gaussian random variable with a standard deviation between 4 and 12 dB, describing the indoor shadowing effect [8][9].

3.3 MULTIPATH PROPAGATION

Due to multipath propagation, a transmitted signal arrives at the receiver not along a single propagation path, but along multiple propagation paths, each with their own specific delay, attenuation, and phase shift, which are function of the traversed electrical path length. Multipath propagation arises from reflection, diffraction, and scattering of the electromagnetic waves on various obstacles in the BFWA radio channel, such as hills, buildings, and billboards [2]. Reflection occurs when an electromagnetic wave impinges upon a smooth surface with very large dimensions compared to the radio frequency (RF) wavelength λ. Diffraction happens when an impenetrable object, with large dimensions compared to λ, obstructs the radio path between transmitter and receiver, causing secondary waves to be formed behind the object. When an electromagnetic wave impinges upon a rough surface, or a surface, whose dimensions are on the order of λ, the energy is scattered in many different directions. For the BFWA situation, reflections are the dominant source of multipath propagation.

As a consequence of this multipath propagation a single impulse transmitted over the BFWA radio channel, appears as a train of attenuated and delayed pulses at the receiver. In this section, we study this delay dispersion more in detail. Subsection 3.3.1 proposes a stochastic model for a multipath channel. Subsection 3.3.2 studies the delay-dispersive effect of the BFWA channel in the delay domain and Subsection 3.3.3 investigates it in the frequency domain.

3.3.1 Delay dispersion

Extensive measurement campaigns have demonstrated that the wireless radio channel can be realistically modelled with a finite length tapped delay line. Different variants of this model have been proposed in the literature and have been adopted by the IEEE 802.16 standardization committee as reference channels [1]. The multipath channel can be modelled by a complex-valued impulse response:

$$h(\tau,t) = \sum_{l=0}^{N_h-1} \underbrace{a_l e^{j\theta_l(t)}}_{A_l(t)} \delta(\tau - \tau_l), \qquad (3.13)$$

where N_h represents the number of taps, a_l is the lth real tap gain, $\theta_l(t)$ is the lth tap's time-varying phase offset, $A_l(t)$ is the lth time-varying complex tap coefficient, τ_l is the lth tap delay, and $\delta(.)$ represents the Dirac function. As will be explained in more detail in Section 3.4, the temporal variation of the channel is mainly caused by the time-varying phase offset. Furthermore, the channel taps are mostly uniformly spaced with a tap spacing, which equals the inverse of the channel sampling rate.

When the number of physical propagation paths that contribute to a certain channel tap becomes very large, $A_l(t)$ will approach a complex Gaussian random variable, by virtue of the central limit theorem. In this case, the phase offset, $\theta_l(t)$, is uniformly distributed, whereas the real tap gain, a_l, is Rayleigh distributed according to the following probability density function:

$$P_A(a) = \frac{2a}{\sigma^2} \exp\left(-\frac{a^2}{\sigma^2}\right), a \geq 0. \qquad (3.14)$$

Delay dispersion refers to the instantaneous behaviour of the channel's impulse response with respect to the time delay τ. The delay-dispersive effect of the channel is characterized by the multipath delay spread in the delay domain and the channel coherence bandwidth in the frequency domain.

3.3.2 Delay domain view of delay dispersion

Delay dispersion can be viewed in the delay domain by means of the power delay profile of the channel, which shows the average received power for a single transmitted pulse as a function of the time delay τ. The time delay τ often refers to the excess delay relative to the first signal arrival at the receiver. The power delay profile is given by:

$$P(\tau) = E\left[\left|h(\tau,t)\right|^2\right] = \sum_{l=0}^{N_h-1} \sigma_{A_l}^2 \delta(\tau - \tau_l),$$

(3.15)

where $\sigma_{A_l}^2 = E\left[a_l^2\right]$ represents the average power of the lth tap.

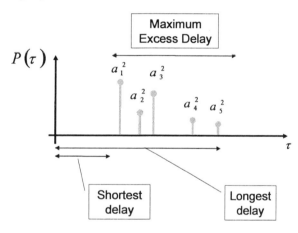

Figure 3.2. Maximum excess delay of a propagation channel.

The amount of delay dispersion in a wireless channel is indicated by the multipath delay spread of the channel. The most commonly used parameters to quantify the multipath delay spread are the maximum excess delay, the average excess delay, and the Root Mean Square (RMS) delay spread. The maximum excess delay, τ_{max}, is defined as the delay of the latest multipath component, whose power remains above a certain threshold (see *Figure 3.2*). The average excess delay is defined as the first order moment of the power delay profile, and is given by:

$$\bar{\tau} = \dfrac{\displaystyle\int_{-\infty}^{\infty} P(\tau)\tau\, d\tau}{\displaystyle\int_{-\infty}^{\infty} P(\tau)\, d\tau} = \dfrac{\displaystyle\sum_{l=0}^{N_h-1} \sigma_{A_l}^2 \tau_l}{\displaystyle\sum_{l=0}^{N_h-1} \sigma_{A_l}^2}. \tag{3.16}$$

The RMS delay spread τ_{RMS} is defined as the square root of the second central moment of the power delay profile, and is given by:

$$\tau_{RMS} = \sqrt{\dfrac{\displaystyle\int_{-\infty}^{\infty} P(\tau)(\tau-\bar{\tau})^2\, d\tau}{\displaystyle\int_{-\infty}^{\infty} P(\tau)\, d\tau}} = \sqrt{\dfrac{\displaystyle\sum_{l=0}^{N_h-1} \sigma_{A_l}^2 (\tau_l-\bar{\tau})^2}{\displaystyle\sum_{l=0}^{N_h-1} \sigma_{A_l}^2}}. \tag{3.17}$$

The RMS delay spread has a fundamental impact on the design of a wireless system. The relationship between the RMS delay spread, τ_{RMS}, and the symbol period, T_s, determines two different degradation categories, frequency-flat fading and frequency-selective fading. On the one hand, a wireless channel is referred to as frequency-flat fading, if the RMS delay spread of the channel is sufficiently small compared to the symbol period ($\tau_{RMS} \ll T_s$), such that the received multipath components of a symbol arrive within one symbol period. Consequently, there is no channel-induced inter-symbol interference (ISI), and no equalization is required at the receiver. A practical rule of thumb for the maximum data rate that can be reliably transmitted over a channel without incorporating an equalizer at the receiver is one-tenth of the inverse of the multipath delay spread. On the other hand, a wireless channel is said to exhibit frequency-selective fading, if the RMS delay spread is large compared to the symbol period ($\tau_{RMS} \gg T_s$), such that the received multipath components of a symbol extend beyond the symbol period. In this case, a complex equalizer is required at the receiver to cope with the resulting channel-induced ISI.

In this book, we rely on the Stanford University Interim (SUI) models, which have been standardized to evaluate candidate air interfaces for BFWA systems. Instead of constructing propagation models for all possible BFWA operating environments, a smaller set of test environments is defined, which adequately span the overall range of possible propagation condidtions. Each operating environment is characterized by a corresponding stochastic model, which is based on the tapped delay line model. Six different channels have been standardized, whose main features are summarized in *Table 3.2.*

Table 3.2. Stanford University Interim channel models.

Model	Terrain	τ_{RMS} (µs)
SUI-1	Best	0.111
SUI-2	Best	0.202
SUI-3	Intermediate	0.264
SUI-4	Intermediate	1.257
SUI-5	Worst	2.842
SUI-6	Worst	5.240

3.3.3 Frequency domain view of delay dispersion

A dual but complementary characterization of delay dispersion can be viewed in the frequency domain. By taking the Fourier transform of the time domain impulse response, we obtain the frequency response of the channel:

$$H(f,t) = \int_{-\infty}^{\infty} h(\tau,t)\exp(-j2\pi f\tau)d\tau = \sum_{l=0}^{N_h-1} \underbrace{a_l e^{j\theta_l(t)}}_{A_l(t)} \exp(-j2\pi f\tau_l) \qquad (3.18)$$

where f is the frequency variable. As can be observed in *Figure 3.3*, the multipath propagation channel results for large bandwidths in frequency selective fading.

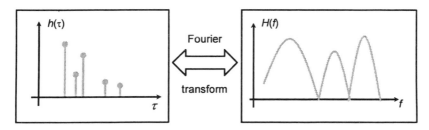

Figure 3.3. Delay dispersion results in frequency selective fading.

The frequency-dependent behaviour of the channel is described by the spaced-frequency correlation function, which represents the correlation between a channel's frequency response and a version shifted over a frequency separation Δf. It is defined as follows:

$$R_A(\Delta f) = E\left[H(f,t)H^*(f+\Delta f,t)\right] = \int_{-\infty}^{\infty} P(\tau)\exp(-2\pi\Delta f\tau)d\tau \qquad (3.19)$$

Naturally, as the frequency separation increases, the correlation will decrease. The degree of frequency-selectivity is indicated by the coherence bandwidth of the channel, which is a statistical measure of the range of frequencies, over which the channel passes all spectral components with approximately equal gain and linear phase. The coherence bandwidth, B_{coh}, is usually defined as the frequency separation, over which the correlation has decreased by 3 dB. Not surprisingly, the spaced-frequency correlation function is the Fourier transform of the power delay profile. Therefore, the coherence bandwidth is inversely proportional to the RMS delay spread, that is, $B_{coh} \cdot \tau_{RMS} = \kappa_1$. For an exponentially distributed power delay profile, $\kappa_1 = 1/(2\pi)$.

Like the RMS delay spread, the coherence bandwidth has a significant impact on the design of a wireless system. The relationship between the coherence bandwidth B_{coh}, and the information bandwidth B_i, distinguished between the two different degradation categories, i.e., frequency-flat versus frequency-selective fading, in a dual but complementary way. If the information bandwidth is sufficiently small compared to the coherence bandwidth ($B_i << B_{coh}$), a wireless channel is said to exhibit frequency-flat fading. Consequently, the different spectral components of the transmitted signal are similarly affected by the channel, and no equalization is required at the receiver. A practical rule of thumb for the maximum data rate that can be reliably transmitted over the channel without equalization is ten percent of the coherence bandwidth. Otherwise, if the information bandwidth is large compared to the coherence bandwidth ($B_i >> B_{coh}$), a wireless channel is referred to as frequency-selective fading. The different spectral components of the signal are now affected differently by the channel, such that a complex equalizer is required at the receiver.

3.4 TEMPORAL VARIATION

Temporal variation refers to the behaviour of the channel's impulse response with respect to the observation time t. The time-varying nature of the BFWA channel is mainly caused by movements of objects within the channel, although, more generally for mobile channels, the relative motion between the transmitter and the receiver additionally contributes to the temporal variation.

In general, the number of physical propagation paths, their attenuation, phase shift, and time delay are time-varying. However, in order for the number of propagation paths, their attenuation and time delay, to change significantly, large dynamic changes in the propagation environment are required. On the other hand, their phase shift can change dramatically with relatively small motions of the propagation environment. Therefore, the variations of the former (the number of propagation paths, their attenuation, and time delay) take place on a much larger time scale, and, thus, can be neglected with respect to the variation of the latter (their phase shift).

In the following, the temporal variation of the channel is characterized by the channel coherence time in the time domain and the Doppler spread in the Doppler frequency domain. Finally, we report on a measurement campaign to characterize the time variations of a BFWA channel in the presence of high-speed traffic.

3.4.1 Time domain view of temporal variation

Temporal variation can be viewed in the time domain by means of the spaced-time correlation function, which describes the correlation between the channel's response for the lth tap and a delayed version with a time separation Δt. It is defined as:

$$R_l(\Delta t) = E\left[A_l(t) \cdot A_l^*(t + \Delta t)\right] \tag{3.20}$$

Naturally, as the time separation increases, the correlation will decrease. The degree of temporal variation is indicated by the coherence time of the channel, which is a statistical measure of the expected time duration, over which the channel's response is essentially invariant. The coherence time, T_{coh}, is usually defined as the time separation, over which the correlation has decreased by 3 dB. In the above discussion, we have considered the temporal variation of a single tap only. Note, however, that each channel tap has its own coherence time. To be on the safe side, one should consider the tap with the smallest coherence time.

The relationship between the coherence time T_{coh}, and the symbol period T_s, distinguishes between two additional degradation categories, slow fading and fast fading. A wireless channel is referred to as slow fading, if the coherence time of the channel is sufficiently large compared to the symbol period ($T_{coh} >> T_s$). Consequently, the time duration, in which the channel behaves in a correlated manner, is long compared to the time duration of a symbol. Adversely, a wireless channel is referred to as fast fading, if the coherence time is small compared to the symbol period ($T_{coh} << T_s$). In this case, the time duration, in which the channel behaves in a correlated manner, is short compared to the time duration of a symbol.

3.4.2 Doppler frequency domain view of temporal variation

A completely analogous characterization of the temporal variation of the channel can be viewed in the Doppler frequency domain. By taking the Fourier transform of the spaced-time correlation function, we obtain the Doppler power spectrum of the lth tap:

$$P_{h,l}(v) = \int_{-\infty}^{\infty} R_l(\Delta t) \exp(-2\pi v \Delta t) d\Delta t , \qquad (3.21)$$

where v is the Doppler frequency variable.

The amount of spectral broadening and, thus, the degree of temporal variation, is indicated by the Doppler spread, also called fading rate, of the channel f_D, which is the (single-sided) width of the Doppler power spectrum. The Doppler spread and the coherence time are reciprocally related, or $f_D \cdot T_{coh} = \kappa_2$.

The relationship between the Doppler spread, f_D, and the information bandwidth, B_i, differentiates between the two degradation categories due to temporal variation in a dual but complementary way. A channel is referred to as slow fading, if the information bandwidth is much larger than the Doppler spread ($B_i \gg f_D$), such that the spectral spreading remains within the information bandwidth. Adversely, a channel is referred to as fast fading, if the information bandwidth is much smaller than the Doppler spread ($B_i \ll f_D$), such that the spectral spreading extends beyond the information bandwidth. Hence, the Doppler spread sets a lower limit on the information bandwidth, or symbol rate that can be used without suffering from fast fading.

3.4.3 Measurement of temporal variation

This section presents the measurement and statistical analysis of the temporal variation of the BFWA radio channel defined by a fixed link at 3.5 GHz in suburban areas [10]. The analysis provides the required information about the temporal stability of the channel in the presence of fast moving traffic on a motorway in the neighbourhood of the receiver.

As introduced before, the temporal variation that signals suffer in mobile and wireless systems has two different causes: the relative movement between the transmitter and the receiver and the movement of objects within the propagation environment, such as people, trees, traffic, machinery, etc. Therefore, even when transmitter and receiver remain fixed, the movement of the environment gives rise to temporal variation. In the case of mobile systems, which are characterised by a high degree of mobility, the temporal variation of the signal due to the relative motion between transmitter and receiver clearly dominates over the temporal variation due to motion of the environment. However, in the case of fixed wireless applications, such as BFWA, the latter effect is the most important one.

In the open literature, measurements and models dealing with temporal variation due to motion of the mobile terminal are frequently found, for instance in [11]. Also, the influence of the movement of people and machinery in indoor radio scenarios has been measured and modelled for different frequency bands in [12],[13],[14],[15],[16]. But in the case of fixed point to multi-point and BFWA systems, there is clearly a lack of information about this subject. Two works have been published in the open literature [17], [18], in which measurements in the 2 GHz and 2.5 GHz bands, respectively, are reported. In [18], the temporal variation is only discussed very briefly. In [17], although the general conclusion is established that temporal variation is significant but can be represented by slowly time-variant functions, an exhaustive statistical analysis is not reported.

This section is organised as follows. Subsection 3.4.3.1 describes the radio-link sites and environments. Subsection 3.4.3.2 presents the experimental set-up as well as the measurement technique. Subsection 3.4.3.3 draws a general description of the measured data for each site. Finally, Subsection 3.4.3.4 presents a statistical analysis of the measured data.

3.4.3.1 Radio link sites and environment

Measurements have been performed between the transmitter and receiver equipment in a suburban area in the City of Santander (Spain). This area consists of one populated zone, which is characterized by a medium-density of dwellings (houses) of up to three storeys, industrial buildings of up to two storeys, and some farmland. The receiver was located at the Engineering School building (ETSIIT), where the receiver antenna is placed in a balcony about 15 meters above street level. The transmitter was placed in four different locations at a nearby suburban area. A motorway, with two lanes in each direction, with an average speed of the cars of about 80 km/h, and with around 40 cars a minute in the rush hour for the two lanes, crosses the path from the transmitter to the receiver. The map in *Figure 3.4* gives a general overview of the radio link environment on a 1:10000 scale.

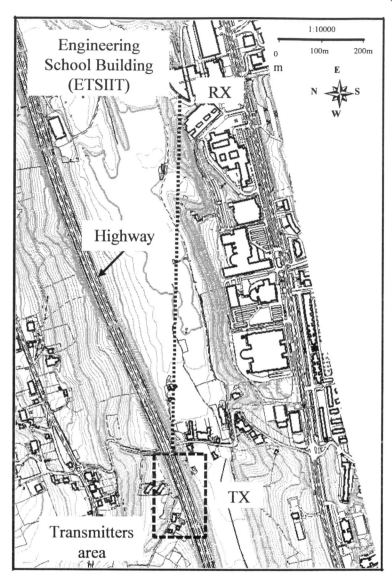

Figure 3.4. General overview of the radio link environment.

The four different transmitter sites are illustrated in *Figure 3.5*. In site LOS1, although LOS, the antenna is approximately at the height of the highway and the direct path is very close to the motorway, such that the traffic strongly influences the signal, and deep fades can be observed, which are clearly correlated with the vehicles. In site NLOS1, the antenna is located near position LOS1, but, in this case, there is no line-of-sight between the transmitter and the receiver, with the obstruction caused by the mound, on which the highway is built. In site LOS2, the antenna is located near the wall of a house further from the receiver and from the motorway, providing the radio-link with a greater clearance, such that the traffic has less influence. In site NLOS2, where a house obstructs the direct ray, the path conditions are very similar to LOS2, and the effects of the traffic are not so large as for NLOS1, although they are measurable.

In the LOS1 and NLOS1 cases, the radio link path is 1000 m long. In the LOS2 and NLOS2 cases, the distance between transmitter and receiver is about 1050 m. In the four cases, the transmitter antenna is attached to a tripod. The antenna height is about 2 meters above the street level, and about two or three meters under the rooftops of the surrounding buildings.

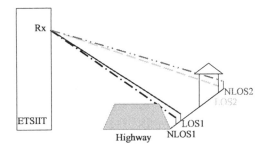

Figure 3.5. Position of the four different transmitter sites.

3.4.3.2 Experimental set-up

The transmitter consists of a signal generator used to generate a carrier at 3.5 GHz with a very stable frequency and a maximum power of 0 dBm. In addition, a power amplifier is employed, with a gain of 30 dB and a 1 dB compression point of +28 dBm. Finally, a medium directive antenna, which is typical for fixed wireless links, launches the signal into the propagation environment. The antenna has a gain of around 15 dB, an azimuth beamwidth of 60°, and an elevation beamwidth of 15°.

The receiver comprises a single antenna with medium gain equal to the transmitter's one, and a block consisting of the spectrum analyser HP70000, the vector signal analyser HP89410A and the downconverter HP89411A. In order to measure and statistically analyse the temporal variation of the radio channel, the received signal has been measured in the time domain with a sample period of 320 ms. The frequency references of all the instruments are locked together. The total observation time at each location is made up of about sixty single 56 s records, in order to observe the signal for a significant period of time. The number of samples in each of these sixty records is around 175.000.

3.4.3.3 Description of measured data

In this subsection, the obtained measurement data are described for the four different transmitter sites, introduced above: LOS1, NLOS1, LOS2, and NLOS2.

A typical result of data recorded in the LOS1 site is shown in *Figure 3.6*. The presence of two types of fluctuations can be observed: continuous fluctuations and deep fades. On the one hand, the scattering from the traffic, single or multiple reflections and diffractions, which are added to the direct path, cause continuous fluctuations around the mean with a variation range of about 2 dB. These contributions suffer Doppler shift, but a relatively narrow Doppler spread is observed. On the other hand, deep fades occur in bursts of several seconds with a dynamic range of around 25 dB. The probability of occurrence of such bursts depends on the intensity of the heavy traffic, and their duration and depth are related with the velocity and size respectively. During the measurement sessions, it has been observed that the deeper fades are clearly correlated with the passing of tall vehicles, like vans, trucks, articulated lorries and buses (coined "heavy traffic" from now on) that obstruct the direct path.

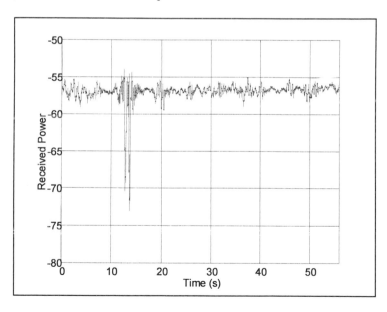

Figure 3.6. Received power over time at the LOS1 site.

As in the LOS1 site, in the NLOS1 site, the presence of two types of fluctuations can be observed, which is shown in *Figure 3.7*. Although the origin and the resulting performance are similar, some differences can be pointed out with respect to the LOS1 site.

As in the LOS1 case, the contributions due to the scattering from the traffic are added to the main diffracted path, resulting in continuous fluctuations. However, a lower level in the fluctuations is observed in this case with respect to LOS1. This is due to the fact that the direct path is completely obstructed in this site, and, therefore, the main contribution to the signal at the receiver is due to a diffracted path from the edge of the mound of the motorway. In this perspective, it should be noted that the transmitter antenna is under the level of the motorway making the arrival of multipath from the traffic more difficult.

Similar to the LOS1 case, the deep fades are now due to the passing of tall vehicles that obstruct the diffracted path but in contrast, these fades occur in bursts of several seconds with a dynamic range up to 50 dB. The probability of occurrence of such bursts depends on the intensity of the traffic, but in this case, the passing of normal cars produce fades with a dynamic range of 10 dB.

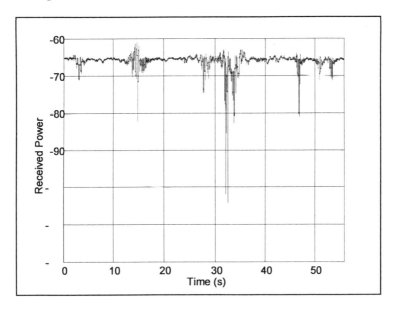

Figure 3.7. Received power over time at the NLOS1 site.

In the LOS2 case, shown in *Figure 3.8*, the clearance of the direct path with respect to the motorway mound is higher that in the previous cases. Most of the time the signal presents very small fluctuations with a dynamic range of about 0.5 dB. Only the passing of heavy traffic gives rise to bursts with a dynamic range of about 2.5 dB. The main contribution to the signal is due to the direct path, such that the scattered signal from the traffic has a minor influence.

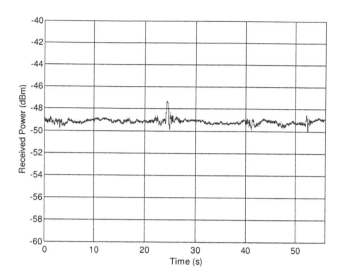

Figure 3.8. Received power over time at the LOS2 site.

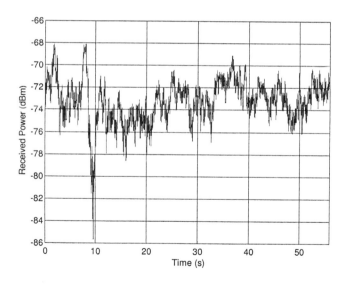

Figure 3.9. Received power over time at the NLOS2 site.

The diffracted path from the wedge of the nearest house is the main contribution to the total signal at the NLOS2 receiver. This is illustrated in *Figure 3.9.* Despite being the main contribution, this signal has a low level, since it is clearly influenced by the multipath signals arriving from different sources in the environment, some of them scattered from the traffic. The fluctuations appear continuously over time with a dynamic range of 8 dB. Some bursts appear intermittently, with fades up to 15 dB.

3.4.3.4 Statistical analysis

Once the measurements have been performed and the corresponding data files have been obtained, the information has been further processed in order to characterise the coherence time of the channel for each of the situations (LOS1, NLOS1, LOS2, NLOS2).

From inspection of the measured data, it can be observed that the envelope variations are not strictly stationary random processes. Different modes of variations can be observed. The explanation of this fact is related to the physical origin of these variations. Two sources can be clearly identified: the first one being the direct obstruction of the main path, direct or diffracted, causing deep but relatively slowly time-varying fades; the second one being the multipath signal arriving at the transmitter and scattered from the nearest environment, including the traffic, and, therefore, subject to the Doppler effect. These latter variations are slightly faster that the former ones. In addition, both types of variations change their characteristics depending on the type and velocity of the vehicles.

The proposed method for further processing these signals is to obtain the major statistical parameters for temporal windows, and then calculate main values (estimates) for the correlation coefficient function. Similar solutions have been adopted by other authors to analyse temporal variation of radio signals [13], [14]. In our case, after several trials with different sizes for the temporal window, the option we took was to fix them according to the time resulting from removing the calm period from each of the 56 seconds long data records.

In the following, the different steps to analyse the obtained measurements in a statistical way are explained. First of all, data is filtered to remove mean variations, which can affect the statistical analysis of data. To eliminate these mean variations, a temporal averaging window was used to obtain the local, or mean, power. Afterwards, the received power was normalised to this mean power.

As can be observed in *Figure 3.6* through *Figure 3.9*, the fades generally occur in bursts, which are associated to the passing of vehicles on the motorway. These fades are separated by periods, in which the signal remains practically constant. Since the parts of the signal that require characterisation are, precisely, such bursts, the quiescent periods between fading bursts were omitted in the analysis. The algorithm used to detect the fading bursts consisted of using a sliding temporal window, inside which the variance of the corresponding signal samples was estimated. In this way, every sample has an associated 'local variance'. Values of the variance above a given threshold were associated to the existence of fading, while values of the variance below that threshold were assumed to indicate a non-fading condition. After several trials with different temporal sizes for the sliding window, a three-second window turned out to be an appropriate one. As for the threshold, several values were also studied, finally deciding on a threshold of 0.2 dB.

Once the bursts were identified, the spaced-time correlation function of the amplitude of the measured signal is calculated[12] [14], in a similar way as equation (3.20). The time separation Δt, after which the correlation function falls below X%, is denominated the X % coherence time.

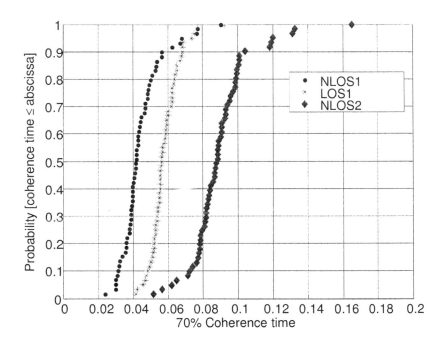

Figure 3.10. PDF of the 70 % coherence time for the the LOS1, NLOS1, and NLOS2 cases.

Figure 3.10 presents the PDF for the 70% coherence time. For the LOS1 case, the mean value of the coherence time is 58 ms, and the probability of having coherence times of less than 40 ms is negligible. For the NLOS1 case, it can be observed that coherence times of less than 20 ms have a negligible probability of occurrence. The mean value of the coherence time for the NLOS1site is 44 ms. Finally, for the NLOS2 case, the mean value of the coherence time is 90 ms, and the probability of finding values below 50 ms is negligible. Position LOS2 is considered as a reference case. In this position, the influence of traffic on the channel is very low and can be considered negligible.

3.5 MULTIPLE ANTENNA CHANNELS

In the previous sections, we have studied the propagation characteristics for the Single-Input Single-Output (SISO) channels, which refers to a system with a single antenna at both ends of the wireless link. However, for BFWA deployments, Multi-Input Multi-Output (MIMO) systems are of particular importance. MIMO wireless systems make use of multiple antennas, or, equivalently, multi-element antennas, at both the transmitter and the receiver, as sketched in *Figure 3.11*. Each transmit-receive antenna pair defines a Single-Input Single-Output (SISO) channel. Thus, the MIMO wireless transmission channel, which is defined between the M_T transmit antennas and the M_R receive antennas, consists of $M_T M_R$ different SISO channels.

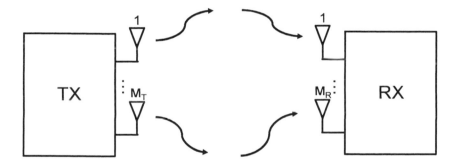

Figure 3.11. The MIMO wireless channel.

BFWA MIMO wireless channel modelling has recently received a great deal of attention. This has resulted in various and numerous channel models, which can be classified in several ways: deterministic versus stochastic, physical versus non-physical, measurements-based versus scatterers-based. An exhaustive comparative study of state-of-the-art outdoor MIMO channel models and their classifications can be found in [21]. For the purpose of positioning the standardized MIMO channel model, in the context of fixed wireless applications, we rather follow the last classification approach.

Measurements-based models refer to the modelling approach that measures a set of real MIMO channel responses through field measurements. From the recorded data, the important characteristics of the MIMO channel are extracted, based on which a model is devised that has similar characteristics. Models based on field measurements were reported in [22], [23], [24], [25], [26]. The alternative scatterers-based models instead postulate a spatial distribution of the scatterers, and then extract the resulting MIMO channel characteristics. Scatterers-based models can be found in [27], [28]. The ability of scatterers-based models to predict MIMO wireless channels critically depends on how realistic their assumed scatterers geometry is. This geometry dependency may entail a large number of parameters and increased modelling complexity. Consequently, measurements-based models are often preferred, which make abstraction of the scattering geometry, and rather capture its effects using a limited set of parameters with given statistics. In this perspective, the IEEE 802.16 Broadband Wireless Access Working Group has considered a measurements-based statistical model for MIMO-based fixed wireless access channels [1], which we have consistently used throughout this book.

This section is organized as follows. Subsection 3.5.1 proposes a stochastic tapped delay line model for the MIMO channel. Subsection 3.5.2 studies the angle dispersion of the MIMO channel, both in the angle and the spatial domain. Finally, Subsection 3.5.3 links spatial correlation to angle dispersion.

3.5.1 Stochastic MIMO channel model

Capitalizing on the SISO channel characterization of Subsection 3.3.1, the space-time impulse response of the BFWA MIMO channel can be modelled by a multi-dimensional time-dependent finite impulse response (FIR) filter. At time t, the equivalent baseband impulse response $\mathbf{H}(\tau,t)$ as a function of the excess delay τ, is given by:

$$\mathbf{H}(\tau,t) = \sum_{l=0}^{N_h-1} \mathbf{H}_l(t)\delta(\tau - \tau_l), \tag{3.22}$$

where the $M_R \times M_T$ complex channel gain matrix, $\mathbf{H}_l(t)$, is the superposition of a large number of propagation paths or rays, but which all arrive at excess delay τ_l. $\mathbf{H}_l(t)$ is the relevant parameter to characterize the MIMO channel fading behaviour. In [25], it was shown to consist of a constant LOS matrix, $\mathbf{H}_l^f$, and a complex Gaussian-distributed NLOS matrix, $\mathbf{H}_l^v$:

$$\mathbf{H}_l = \sqrt{P_l}\left(\sqrt{\frac{K_l}{K_l+1}}\mathbf{H}_l^f + \sqrt{\frac{1}{K_l+1}}\mathbf{H}_l^v\right), \tag{3.23}$$

where P_l is the average power of the lth channel tap, and is determined by the power delay profile of (3.15). K_l is the Ricean K-factor of the lth tap, which can be non-zero only for the first tap. The other taps, having at least experienced one reflection, can only be NLOS. The deterministic fixed LOS matrix H_l^f simply describes the relative phase shifts between the multiple antennas, at the transmitter and the receiver. The NLOS fading matrix H_l^v has complex Gaussian entries. The Gaussian distribution is justified by the fact that H_l^v is the superposition of a large number of individual ray responses. The joint distribution of entries of H_l^v however, is dependent on the channel's scattering geometry and the resulting angle dispersion. In Subsection 3.5.2, we characterize the angle dispersion phenomenon, and describe its impact on the joint distribution of the NLOS fading matrix H_l^v.

3.5.2 Angle dispersion

During their radio propagation, the transmitted electro-magnetic waves experience multiple reflections, diffractions and scatterings. Consequently, a large number of propagation paths, or, rays, will arrive at the receiver from a wide range of angles. This phenomenon is known as angle dispersion, and can be characterized both in the angle domain and its dual space domain.

3.5.2.1 Angle domain view of angle dispersion

In the angle domain, the angle dispersion effect can be described using the angular power spectrum $P_h(\theta)$, which characterizes the received power on a receiving antenna as a function of the angle of arrival θ as follows

$$P_h(\theta) = \sum_{k=0}^{+\infty} P_k \delta(\theta - \theta_k),$$ (3.24)

where θ_k is the angle of arrival and P_k the received power of the kth propagation path or ray.

The mean arrival angle $\bar{\theta}$ is found as

$$\bar{\theta} = \frac{\sum_{k=0}^{+\infty} P_k \theta_k}{\sum_{k=0}^{+\infty} P_k}.$$ (3.25)

The angular dispersion around the mean angle of arrival $\bar{\theta}$ can be measured by the RMS angle spread σ_θ, defined as

$$\sigma_\theta^2 = \frac{\displaystyle\sum_{k=0}^{+\infty} P_k \left(\theta_k - \bar{\theta}\right)^2}{\displaystyle\sum_{k=0}^{+\infty} P_k}.$$ (3.26)

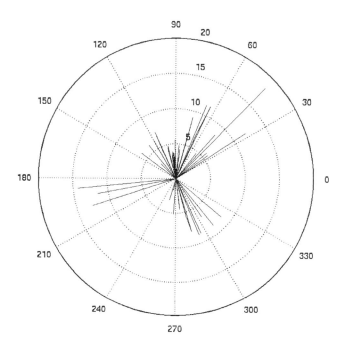

Figure 3.12. Example of an angular power spectrum for indoor propagation.

The angle dispersion profile directly reflects the multipath geometry and the spatial distribution of the scatterers in a given fixed wireless access radio channel. In particular, the severe-multipath and dense-scattering nature of the fixed wireless access medium is observed as a large angle spread and a dense angular power spectrum.

3.5.2.2 Space domain view of angle dispersion

Based on the angular power spectrum, we can derive the spatial correlation $R(\Delta d)$, which measures the resemblance between two observations of the channel at two antennas spaced by a distance Δd. It can be calculated according to

$$R(\Delta d) = \int_{-\pi}^{\pi} P_h(\theta) e^{j\frac{2\pi\Delta d}{\lambda}\sin(\theta)} d\theta, \tag{3.27}$$

where λ denotes the transmission wavelength. The spatial correlation between the two antennas is shown to decay when their spatial separation Δd increases. The spatial separation, over which the spatial correlation has decreased by 3 dB, defines the coherence distance of the channel D_c.

The characterization and modelling of angle dispersion has hardly any relevance for SISO systems. However, it acquires crucial importance for multiple-antenna systems, in the context of which it is essential for describing the channel's spatial characteristics, in particular the spatial correlation on the NLOS fading matrix, H_l^v.

3.5.3 Spatial correlation

A poor scattering and low angle dispersion preserve the spatial correlation between the entries of each individual ray into the entries of H_l^v. Conversely, a rich scattering and large angle dispersion, through summing a large number of independent correlated rays, will cancel any correlation between the entries of H_l^v. In the latter scenario, the NLOS fading matrix has independent identically-distributed (i.i.d.) complex Gaussian-distributed entries.

The spatial complex correlation between transmit antenna i_1 and transmit antenna i_2, associated to the NLOS component of lth channel tap H_l^v, is measured by

$$\rho_{i_1 i_2}^{Tx,l} = E\left[H_l^v[j,i_1]\left(H_l^v[j,i_2]\right)^*\right] \quad 1 \le j \le M_R. \tag{3.28}$$

In [24], it is assumed that the spatial correlation coefficients at the transmitter $\left\{\rho_{i_1 i_2}^{Tx,l}\right\}_{1 \le i_1 i_2 \le M_T}$, are independent of the receive antenna index j. This is due to the fact that the M_R receive antennas view the same surrounding scatterers and, consequently, cause the same angular power spectrum at the transmitter $P_{H_l^v}^{Tx}(\theta)$, which determines the spatial correlation coefficient in (3.28), as follows [29], [24]:

$$\rho_{i_1 i_2}^{Tx,l} = \int_{-\pi}^{\pi} P_{H_l^v}^{Tx}(\theta) e^{j\frac{2\pi d}{\lambda}\sin(\theta)} d\theta, \tag{3.29}$$

where d stands for the antenna spacing. Analogously, the spatial complex correlation between receive antenna j_1 and receive antenna j_2, associated to the NLOS component of the lth channel tap H_1^v, is defined by $\rho_{j_1 j_2}^{Rx,l}$ and is similarly assumed to be independent of the transmit antenna index i. Given these complex correlation coefficients, we can define the following Hermitian spatial channel covariance matrices [30], [24]:

$$
\mathbf{R}_{Tx}^1 = \begin{bmatrix} \rho_{11}^{Tx,l} & \cdots & \rho_{1M_T}^{Tx,l} \\ \vdots & \ddots & \vdots \\ \left(\rho_{1M_T}^{Tx,l}\right)^* & \cdots & \rho_{M_T M_T}^{Tx,l} \end{bmatrix},
$$

$$
\mathbf{R}_{Rx}^1 = \begin{bmatrix} \rho_{11}^{Rx,l} & \cdots & \rho_{1M_R}^{Rx,l} \\ \vdots & \ddots & \vdots \\ \left(\rho_{1M_R}^{Rx,l}\right)^* & \cdots & \rho_{M_R M_R}^{Rx,l} \end{bmatrix}
$$

(3.30)

which completely describe, for the lth channel tap, the correlations between the transmit antenna elements, on the one hand, and the receive antenna elements, on the other hand. However, to fully describe the correlation properties of H_1^v we need to characterize the correlation between two arbitrary transmission coefficients connecting two different sets of antennas, which is defined by the correlation coefficient $\rho_{i_1 i_2, j_1 j_2}^l$ which was shown in [24], under the assumption that the transmit and receive correlations are independent, to be equivalent to

$$
\rho_{i_1 i_2, j_1 j_2}^l = \rho_{i_1 i_2}^{Tx,l} \cdot \rho_{j_1 j_2}^{Rx,l}.
$$

(3.31)

Based on this relation, a simple formula to generate correlated MIMO channel instances H_1^v starting from i.i.d. MIMO channel realizations H_1^{iid} can be defined as follows:

$$
H_1^v = \left(\mathbf{R}_{Rx}^1\right)^{1/2} H_1^{iid} \left(\mathbf{R}_{Tx}^1\right)^{1/2}.
$$

(3.32)

3.6 SUMMARY

In this chapter, we have reviewed the characteristics and the main impairments of the BFWA radio channel, as it imposes an important boundary condition on BFWA radio systems. As is the case for any radio channel, the BFWA radio channel manifests itself through two types of fading impairments: large-scale and small-scale fading.

From the large-scale fading perspective, the BFWA radio channel exhibits an outdoor path loss behaviour, which differs significantly from that of the highly idealized free-space propagation channel. Furthermore, for a CPE that is located indoor, the outdoor-to-indoor penetration loss has to be taken into account as well, with practical values well above 15 dB.

From the small-scale fading perspective, the BFWA radio channel features considerable multipath propagation, in combination with limited temporal variation, causing signal energy dispersion along three dimensions. These are the delay, the Doppler frequency, and the angle dimension.

Delay dispersion is quantified by the channel's delay spread in the delay domain, with typical values for the BFWA radio channel ranging between 0.111 µs and 5.240 µs, depending on the terrain conditions. When viewed in the frequency domain, delay dispersion induces frequency selectivity, which is quantified by the channel's coherence bandwidth. Since the signal bandwidth extends well beyond the coherence bandwidth, BFWA radio systems have to cope with frequency-selective fading.

Doppler shift dispersion, which is quantified by the Doppler spread in the Doppler frequency domain, is mainly caused by moving objects in the propagation environment, in addition to multipath propagation. When viewed in the time domain, Doppler shift dispersion induces time variations in the channel, which are quantified by the channel's coherence time. As derived from an extensive measurement campaign, typical mean values of the 70 % coherence time vary between 44 ms and 90 ms. Because the symbol period (as well as the frame length) remains well below the coherence time, BFWA radio systems are characterized by slow fading conditions.

Finally, angle dispersion, which is especially relevant to multiple antenna systems, is quantified by the angle spread in the angle domain. When viewed in the spatial domain, angle dispersion induces spatial selectivity, which is quantified by the coherence distance.

3.7 REFERENCES

[1] IEEE 802.16 Broadband Wireless Access Working Group, "Channel Models for Fixed Wireless Applications", Revision 4.0, IEEE802.16.3c-01/29r4, July 2001.

[2] J.B. Andersen, T.S. Rappaport, S. Yoshida, "Propagation Measurements and Models for Wireless Communication Channels", IEEE Communications Magazine, Vol. 33, No. 1, pp. 42-49, January 1995.

[3] B. Sklar, "Rayleigh Fading Channels in Mobile Digital Communication Systems – Part I: Characterization", IEEE Communications Magazine, Vol. 35, No. 7, pp. 90-100, July 1997.

[4] A. J. Paulraj, C.B. Papadias, "Space-Time Processing for Wireless Communications", IEEE Signal Processing Magazine, Vol. 14, No. 6, pp. 49-83, November 1997.

[5] C.L. Hong, I.J. Wassell, G.E. Athanasiadou, S. Greaves, M. Sellars, "Wideband Channel Measurements and Characterisation for Broadband Wireless Access", Twelfth IEE International Conference on Antennas and Propagation (ICAP), Vol. 1, pp. 429-432, April 2003.

[6] V. Erceg, L.J. Greenstein, S.Y. Tjandra, S.R. Parkoff, A. Gupta, B. Kulic, A.A. Julius, R. Bianchi, "An Empirically Based Path Loss Model for Wireless Channels in Suburban Environments", IEEE Journal on Selected Areas in Communications, Vol. 17, No. 7, July 1999.

[7] Y. Miura, Y. Oda, T. Taga, "Outdoor-to-indoor Propagation Modelling with the Identification of Path Passing Through Wall Openings", PIMRC, Vol. 1, pp. 130-134, September 2002.

[8] H. Hashemi, "The Indoor Radio Propagation Channel", Proceedings of IEEE, Vol. 81, No. 7, pp. 943-968, July 1993.

[9] T.S. Rappaport, "Wireless Communications: Principles and Practice", Prentice Hall, 2nd Edition, 2002.

[10] B. Cobo, S. Loredo, D. Mavares, F. Medina, R.P. Torres, M. Engels, "Measurement and Statistical Analysis of the Temporal Variations of a Fixed Wireless Link at 3.5 GHz", accepted for publication in Wireless Personal Communications, 2006.

[11] D.C. Cox, "Delay Doppler Characteristics of Multipath Propagation at 910 MHz in Suburban Mobile Radio Environment", IEEE Transactions on Antennas and Propagation, Vol. 20, No. 5, pp. 625-635, 1992.

[12] H. Hashemi, M. Mcguire, T. Vlasschaert, D. Tholl, "Measurements and Modelling of Temporal Variations of the Indoor Radio Propagation Channel", IEEE Transactions on Vehicular Technology, Vol. 43, No. 3, pp. 733-737, August 1994.

[13] R.J.C. Bultitude, "Measurements, Characterisation and Modelling of Indoor 800/900 MHz Radio Channels for Digital Communications", IEEE Communications Magazine, Vol. 25, pp. 5-12, 1987.

[14] P. Marinier, G.Y. Delisle, C.L. Despins, "Temporal Variations of the Indoor Wireless Millimeter-Wave Channel", IEEE Transactions on Antennas and Propagation, Vol. 46, No. 6, pp. 928-934, June 1998.

[15] S. Loredo, R.P. Torres, "Experimental Analysis of Temporal Variations in Indoor Radio Channels at 1.8 GHz", Microwave and Optical Technology Letters, Vol. 35, No. 2, pp. 132-137, October 2002.

[16] S. Thoen, L. Van der Perre, M. Engels, "Modeling the Channel Time-Variance for Fixed Wireless Communications", IEEE Communications Letters, Vol. 6, No. 8, pp. 331-333, August 2002.

[17] W. Mohr, "Radio Propagation for Local Loop Applications at 2 GHz", Third Annual International Conference on Universal Personal Communications, pp. 119-123, September 1994.

[18] M.J. Gans, N. Amatay, Y.S. Yeh, T.C. Damen, R.A.Valenzuela, C. Cheon, and J. Lee, "Propagation Measurements for Fixed Wireless Loops (FWL) in a Suburban Region with Foliage and Terrain Blockages", IEEE Transactions on Wireless Communications,Vol. 1, No. 2, pp. 302-310, February 2002.

[19] Agilent Product Note 89400-9, "Downconverted Measurements Using the HP 89410A and HP 89411A".

[20] Agilent Product Note 89400-10, "Time-Capture Capabilities of the Agilent 89400 Series Vector Analyzers".

[21] K. Yu, "Multiple-Input Multiple-Output Radio Propagation Channel: Characteristics and Models", PhD Thesis, Royal Institute of Technology, Stockholm, Sweden, January 2005.

[22] K.I. Pedersen, J.B. Andersen, J.P. Kermoal, P. Mogensen, "A Stochastic Multiple-Input-Multiple-Output Radio Channel Model for Evaluation of Space-Time Coding Algorithms", IEEE Proceedings of VTC-Fall, Vol. 2, pp. 893-897, September 2000.

[23] C. C. Martin, J. H. Winters, N. R. Sollenberger, "Multiple-Input Multiple-Output (MIMO) Radio Channel Measurements, IEEE Proceedings of VTC-Fall, Vol. 2, pp. 774-779, 2000.

[24] J. P. Kermoal, L. Schumacher, K. L. Pedersen, P. E. Mogensen, F. Frederiksen, "A Stochastic MIMO Radio Channel Model with Experimental Validation", IEEE Journal on Selected Areas in Communications, Vol. 20, No. 6, pp. 1211-1226, August 2002.

[25] P. Soma, D. S. Baum, V. Erceg, R. Krishnamoorthy, A. J. Paulraj, "Analysis and Modeling of Multiple-Input Multiple-Output (MIMO) Radio Channel Based on Outdoor Measurements Conducted at 2.5 GHz for fixed BWA applications", IEEE Proceedings of ICC, Vol. 1, pp. 272-276, April 2002.

[26] V. Erceg, P. Soma, D. S. Baum, S. Catreux, "Multiple-Input Multiple-Output Fixed Wireless Radio Channel Measurements and Modeling Using Dual-Polarized Antennas at 2.5 GHz", IEEE Transactions on Wireless Communications, Vol. 3, No. 6, pp. 2288-2298, November 2004.

[27] D. Shiu, G. J. Foschini, M. J. Gans, J. M. Kahn, "Fading Correlation and Its Effect on the Capacity of Multielement Antenna Systems", IEEE Transactions on Communications, Vol. 48, No. 3, pp. 502-513, March 2000.

[28] D. Gesbert, H. Boelcskei, D.A. Gore, A.J. Paulraj, "Outdoor MIMO Wireless Channels: Models and Performance Prediction", IEEE Transactions on Communications, Vol. 50, No. 12, pp. 1926-1934, December 2002.

[29] L. Schumacher, K. I. Pedersen, P. E. Mogensen, "From Antenna Spacings to Theoretical Capacities - Guidelines for Simulating MIMO Systems", Proceedings of PIMRC, Vol. 2, pp. 587-592, September 2002.

[30] D. P. McNamara, M. A. Beach, P.N. Fletcher, P. Karlsson, "Initial Investigation of Multiple-Input Multiple-Output Channels in Indoor Environments", IEEE Proceedings of Benelux Chapter Symposium on Communications and Vehicular Technology, pp. 139-143, October 2000.

Chapter 4

Guaranteeing Quality-of-Service
The BFWA Data Link Layer

Marc Engels

4.1 INTRODUCTION

The data link layer, i.e., layer 2 as defined in the 7-layer Open Systems Interconnection (OSI) Reference Model [1], is an essential function of any BFWA solution. Its main function is to share the available communication resources in a fair way between the different connections, taking into account their Quality-of-Service (QoS) requirements. To guarantee QoS, a central scheduling approach is the preferred choice. In such an approach, the central scheduler has an overview of all communication requests, and, hence, can determine the optimal communication order.

The data link layer of the IEEE 802.16 standard [2] pursues this approach. It is common for all physical layers (PHYs) in the standard, and is inspired on the DOCSIS cable modem standard [3]. As a consequence, it contains extensive features for managing QoS, at the cost of an increased implementation complexity, especially at the base station.

Because of its state-of-the-art structure and performance, we will restrict this chapter to the IEEE 802.16 data link layer. After this introduction, Section 4.2 first reviews the history and global structure of the IEEE 802.16 standard. Section 4.3 introduces the notion of service-specific convergence sublayers, which form the interface between the services and the data link layer. Next, Section 4.4 reviews the actual data link layer, while its performance is assessed in Section 4.5. Finally, we will conclude this chapter by summarizing the main observations.

4.2 THE IEEE 802.16 STANDARD

Before discussing the data link layer, we will first review the historical evolution of the IEEE 802.16 standards. The base standard [4] was finalized at the end of 2001, and specified the data link layer, as well as the physical communication mode for Line-of-Sight (LOS) operation in the 10 to 66 GHz frequency bands. In January 2003, the IEEE 802.16a standard [5] that defined additional physical layers for Non-Line-of-Sight (NLOS) operation between 2 and 11 GHz, was completed.

The 802.16 standard is very flexible with a lot of parameters, including carrier frequencies, channel bandwidths, duplex methods, and supported protocols for the convergence layer. Therefore, it was felt useful to define system profiles, which are typical implementation cases with their list of features. These system profiles will form the basis for conformance testing. For the base standard, detailed system profiles for operation between 20 and 66 GHz were defined in IEEE 802.16c-2002 [6].

The 802.16d Task Group initially intended to define similar system profiles for the IEEE 802.16a standard with carrier frequencies between 2 and 11 GHz, but finally ended up realizing a converged standard, IEEE 802.16-2004 [7], in mid 2004. In the remainder of the book, we will use this standard version as a reference.

Currently, two task groups just finalized further extensions of the 802.16 standards:

- The 802.16e Task Group extends the standard towards slow mobility, up to 70 kilometers per hour. This requires, among others, the definition of a handover mechanism between the base stations. The amendment has been approved on 7 December 2005.
- The network management Task Group defines the Management Information Base (MIB) with the associated management plane procedures and services. These standard amendments are called 802.16f and 802.16g, respectively. The 802.16f amendment has been approved as a standard on 22 September 2005. The definition of a new standard amendment, 802.16i, for a mobile management information base was recently initiated.

The IEEE 802.16 standard combines a large set of modes and options, which are mostly incompatible. An overview of these modes is given in *Table 4.1.*

The 10 to 66 GHz communication mode is called WirelessMAN-SC. It features the basic data link layer, and can be applied with time division or frequency division duplex. Where Frequency Division Duplex (FDD) assumes a fixed ratio between up- and downlink communication, Time

Division Duplex (TDD) is more flexible at the cost of potential base station to base station interference problems [8]. To counteract this interference, a fixed time portion of the TDD frame can be allocated to up- and down-link transmission. However, in this way, also the flexibility will be lost. Therefore, a dynamic TDD allocation is more often used, in combination with an intelligent time-slot allocation method.

Table 4.1. The IEEE 802.16 standard combines multiple (incompatible) modes.

Name	Applicability	PHY	Data Link	Duplex
WirelessMAN-SC	Licensed 10-66 GHz	Single carrier	Basic	TDD, FDD
WirelessMAN-SCa	Licensed below 11 GHz	Single carrier, (STC)	Basic, (ARQ), (AAS)	TDD, FDD
WirelessMAN-OFDM	Licensed below 11 GHz	OFDM, (STC)	Basic, (ARQ), (AAS), (MESH)	TDD, FDD
WirelessMAN-OFDMA	Licensed below 11 GHz	OFDMA, (STC)	Basic, (ARQ), (AAS)	TDD, FDD
WirelessHUMAN-SCa	Unlicensed below 11 GHz	Single carrier	Basic + DFS, (ARQ), (AAS)	TDD
WirelessHUMAN-OFDM	Unlicensed below 11 GHz	OFDM (STC)	Basic + DFS, (ARQ), (AAS), (MESH)	TDD
WirelessHUMAN-OFDMA	Unlicensed below 11 GHz	OFDMA, (STC)	Basic + DFS, (ARQ), (AAS)	TDD

Depending on the physical layer (PHY) communication scheme, three modes are defined for licensed frequency bands below 11 GHz: Single Carrier (WirelessMAN-SCa), Orthogonal Frequency Division Multiplexing (WirelessMAN-OFDM), and Orthogonal Frequency Division Multiple Access (WirelessMAN-OFDMA). All three modes can work in TDD or FDD configuration. On top of the basic functionality, extra optional features, like Automatic Retransmission reQuest (ARQ), Space-Time Coding (STC), and Adaptive Antenna Systems (AAS) are defined. For the WirelessMAN-OFDM also an optional Mesh communication scheme is foreseen.

For the unlicensed frequency bands below 11 GHz, the mode is called WirelessHUMAN. Because again the three physical communication schemes are used, we can better indicate them with WirelessHUMAN-SCa, WirelessHUMAN-OFDM, and WirelessHUMAN-OFDMA, respectively. In these unlicensed frequency bands only TDD is allowed. On top of the basic data link layer, Dynamic Frequency Selection (DFS) is mandatory for the license-exempt frequency bands. Moreover, the same additional options as above are foreseen.

The basic data link layer of the IEEE 802.16 standard assumes a topology with a central base station. The subscribers are fixed, and connected to this base station. The base station schedules the up- and down-link communication for all subscriber stations that are connected to it.

To avoid interference, a frequency planning is needed such that neighboring base stations use different carrier frequencies. The frequency planning is different for the above 10 GHz LOS and the sub-11 GHz NLOS communication. In the former, less frequency bands are required for base stations with directional antennas. For instance, if each base station has four 90° sectors, 4 frequencies suffice for a FDD deployment.

Although for the NLOS communication mode, the directivity of antennas can be less exploited because of possible reflections, sectorized base stations, with up to 6 sectors, are frequently used to increase the capacity of a base station site. The coordination between these sectors is typically limited to time synchronization.

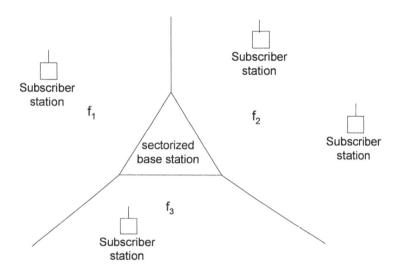

Figure 4.1. The IEEE 802.16 standard is based on a point-to-multipoint topology.

The IEEE 802.16 standard has a layered structure. The data link layer, consisting of the Medium Access Control (MAC) and the security sub-layers, form a common functionality on top of multiple PHY communication modes. The MAC sub-layer provides mechanisms to share the single communication medium over the multiple subscriber stations. The security sub-layer defines the encryption algorithms and the associated protocols for the exchange of the encryption keys.

The interface between the data link layer and the higher protocol layers is formed by a convergence sublayer. Two different service-specific convergence sublayers are currently defined in the standard: one for Asynchronous Transfer Mode (ATM) and one for packet based protocols, like Ethernet or the Internet Protocol (IP). As the convergence sublayers classify the incoming packets according to their QoS requirements, they play a crucial role in guaranteeing those requirements.

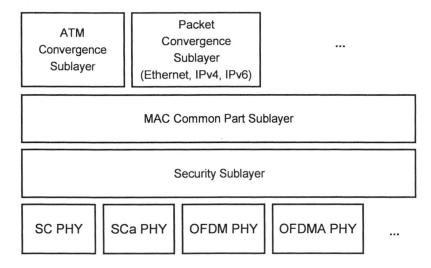

Figure 4.2. The IEEE 802.16 standard has a layered structure.

4.3 CONVERGENCE SUBLAYERS

The main functionality of a convergence sublayer is to compare the parameters of each Service Data Unit (SDU), which is a packet that has to be transmitted, with a set of classifiers. These classifiers will map the packet on one of the N_{oc} open connections, with a specific Connection IDentifier (CID), as illustrated in *Figure 4.3*.

Depending on the selected connection, part of the payload header can be suppressed. By doing so, some header overhead is removed, resulting in a higher effective data rate. The suppressed header is reconstructed at the receiving side.

Each classifier is a rule, with an associated priority, that compares parameters of the packet to be transmitted with a value, or a range of values. Prior to the comparison, a mask can be applied to the parameter. For Ethernet, the applicable parameters include the source and destination MAC address and the user priority field. Similarly, Internet traffic can be classified according to the IP source and destination address, the source and destination ports, and the type of service field. For Internet traffic over Ethernet, any combinations of these fields can be used.

The ATM convergence sublayer supports in a similar way the mapping of virtual channels to connections.

It should be remarked that the many comparisons for the classifiers form a substantial processing load. Therefore, the number of rules and parameters for classifiers are limited in any practical implementation.

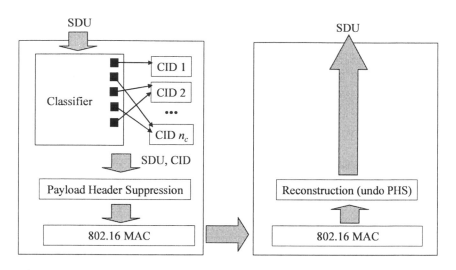

Figure 4.3. The convergence sublayer plays a crucial role in guaranteeing QoS.

To guarantee QoS, each connection is associated to a service flow. A service flow is a description of the QoS parameters of a connection, like the traffic priority, the minimum reserved traffic rate, the maximum traffic burst size, the maximum sustained traffic rate, the maximum latency, the tolerated time jitter, and the scheduling mechanisms. Four types of scheduling mechanisms are supported:

- The Unsolicited Grant Service (UGS) is typically used for applications with constant bit-rate, like voice communications. For UGS scheduling, transmission slots are regularly reserved at the base station. The subscriber station does not need to send transmission requests.
- The Real-Time polling Service (RT-pS) is intended for real-time traffic with variable bit-rate, like compressed video. The base station will regularly poll a specific subscriber station for transmission requests. In addition, the subscriber stations can also piggy back a transmission request on a scheduled uplink transmission.
- For the non Real-Time polling Service (nRT-pS), the base station tries to reserve a minimum data rate to a connection. This mode is typically used for high-throughput services, like the File Transfer Protocol (FTP). A subscriber station can use any of the

mechanisms to send a transmission request: piggy backing, using a contention period, or responding to a multicast polling.

- Finally, the Best Effort (BE) service does not provide any QoS guarantees. All mechanisms for uplink transmission requests can be used.

Remark that the scheduler, which takes into account the service flow requirements and comes up with a transmission order of the SDUs, is another complicated process in the base station. The precise algorithm for this scheduler is vendor dependent and forms an important aspect of product differentiation, both in price and performance, between the equipment manufacturers.

In addition to the QoS parameters, a service flow also has an associated payload header suppression rule. For each service flow, two patterns are defined: the Payload Header Suppression Mask (PHSM) and the Payload Header Suppression Field (PHSF). The operation of payload header suppression is illustrated in *Figure 4.4*. At the transmitter, only the header bits, for which the PHSM equals 0, are transmitted. For the other bits, it is verified that they correspond to the PHSF. At the receiver, the bits of the PHSF arc inserted into the transmitted data stream. The merging order is defined by the PHSM.

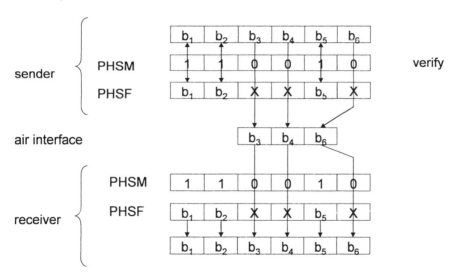

Figure 4.4. Payload header suppression mechanism.

The classifier rules, payload header suppression rules, and service flows can be installed when the equipment is deployed. However, they can also be dynamically downloaded over-the-air on the subscriber stations, after installation of the system. This allows adding new services to already deployed systems.

4.4 DATA LINK LAYER

The data link layer consists of a Medium Access Control (MAC) functionality and a security sub-layer. The common MAC protocol is based on Time-Division Multiple Access (TDMA). The time slots are scheduled centrally by the base station for the multiple users. The transmissions are grouped in frames with a constant length, which is configurable by the base station. Typically, a value of 10 ms is chosen for the frame duration. The downlink and uplink frame structures are shown in *Figure 4.5*. The downlink frame is broadcasted to all terminals and consists of a preamble, and a broadcast control field, followed by bursts to subscriber stations. The bursts to subscriber stations can optionally have an extra preamble, such that the channel estimation can be re-trained in case of mobile terminals. An additional advantage of this preamble is that a subscriber station does not have to listen to parts of the downlink frame that are not intended for him. The format of these downlink bursts are described in the downlink map, which is part of the broadcast control field.

The broadcast control field also contains the uplink map, which describes the uplink frame. The uplink frame consists of uplink time bursts, each starting with a preamble. In a TDD system, the uplink frame is concatenated to the downlink frame. For FDD, they are simultaneously transmitted on two frequencies. The standard allows that FDD subscriber stations can be restricted to half-duplex operation, which largely simplifies their implementation.

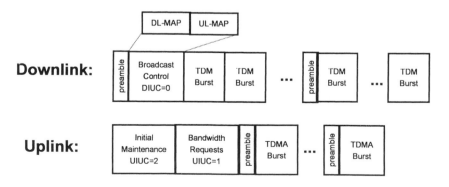

Figure 4.5. Uplink and downlink frame structure of the IEEE 802.16 standard.

To provide the subscriber stations a possibility to enter the network and to request uplink transmission opportunities, some limited time periods are foreseen, in which the subscriber stations compete for the transmission medium and, hence, can collide. In case of a collision, the subscriber stations back-off and wait for a random period before making a new transmission attempt.

On top of this basic MAC mechanism, the protocol contains several features to increase the efficiency of transmission. First, multiple communications can be concatenated into a single up- or downlink burst. The different communications in a burst are distinguished by their CID. The benefit of this concatenation is the reduced overhead of the downlink map, at the cost of additional receiver processing for decoding the complete burst and selecting the relevant communications. Second, the standard supports the fragmentation and packing of communications. This simplifies the filling of the communication frame and, hence, minimizes the required stuffing, and, hence, increases the efficiency of the ARQ algorithm. Third, to support error detection, communications are protected by a Cyclic Redundancy Check (CRC) code.

The entry of a new subscriber station into the network consists of the following 10 steps:

1. First, the subscriber station scans the various carrier frequencies until it detects the downlink frame and has established frequency synchronization, based on the broadcasted preamble, with the base station.
2. Next, the subscriber station decodes the uplink channel description message, which is periodically transmitted by the base station. This message contains the uplink transmission parameters.
3. In the third step, the subscriber station selects an initial maintenance time slot and sends a message to the base station. Based on the response of the base station, it adjusts its transmission timing and power. This initial ranging iteratively continues until timing and power synchronization is achieved.
4. Once the base station and subscriber station are synchronized, the capabilities that are supported by both the subscriber station and the base station, are negotiated.
5. Next, the base station authorizes the subscriber station and exchanges the encryption keys.
6. The subscriber station is also registered.
7. For managed subscriber stations, three additional tasks are performed: the establishment of IP connectivity,
8. the establishment of the time of day,

9. and the transfer of the operational parameters.
10. Finally, three connections are set-up for management purposes:
    - The basic connection has the highest priority and is used for urgent, short messages of the MAC layer.
    - The primary management connection is intended for longer MAC messages, which are less sensitive to delay.
    - The secondary management connection supports higher layer management messages, for instance for the Dynamic Host Configuration Protocol (DHCP), and the Simple Network Management Protocol (SNMP).

Next to the mandatory features of the basic MAC layer, a number of optional features are defined: Automatic Retransmission reQuest (ARQ), Adaptive Antenna Systems (AASs), and a mesh communication mode (MESH).

An ARQ mechanism allows for fast retransmission of data in case of transmission errors. This feature largely improves performance of data transfer applications. As shown on *Figure 4.6*, the ARQ mechanism is a hybrid scheme that combines selective retransmission with roll back. In the selective retransmission mode, an acknowledgement is sent for each of the transmitted blocks. When one of the blocks is not received correctly, this block is retransmitted. In the roll-back mode, only an acknowledgement has to be sent for the last successfully received block. From there on, all blocks are re-transmitted.

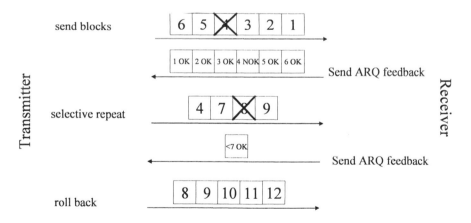

Figure 4.6. The automatic retransmission request mechanism supports both selective retransmission and roll-back.

Although the selective retransmission mechanism is more efficient with respect to transmission bandwidth, it requires a more complicated management and storage architecture.

AASs are intended to extend the range or to increase the capacity of a base station. As shown in *Figure 4.7*, the idea is that multiple antennas are put at the base station site. The different antenna signals can be constructively combined to form highly directional beams, both in up- and down-link. In a more general scheme, multiple beams or full space-time processing is also possible. In addition to range and capacity extension, these directed antenna arrays also suppress interference from other users and cells.

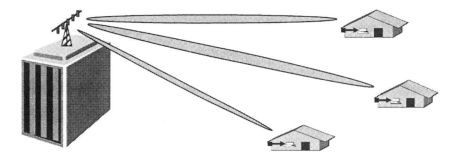

Figure 4.7. Advanced antenna systems extend range and increase capacity.

A separate part of the up- and down-link frame can be reserved for AASs. If the system is only used to increase the capacity, no other changes are required, as the subscriber stations can still receive all management messages that have been transmitted without directing the antennas. Similarly, the base station can easily receive the signal from the subscriber station, even without knowledge about its position.

The increased capacity can be achieved by improvement of the link budget, as well as by parallel transmission to multiple subscriber stations. The latter is called Space Division Multiple Access (SDMA). For both schemes, the base station requires reliable Channel State Information at the transmitter. In an FDD mode, this information is obtained by feeding it back from the subscriber station. For TDD, also an alternative approach exists, in which the reciprocity of the up- and down-link communication channels is exploited. However, this reciprocity and necessary calibration of the radio front-ends, which are an integral part of the communication link, is still a disputed subject, which is actively researched [9].

When the AASs are also used for range extension, the antenna gain by directing the antennas is needed to perform a communication between the subscriber station and the base station. However, at the initial network entry, the base station does not have the necessary information to direct the antennas. Therefore, the standard foresees an active scanning technique, where preambles are transmitted in different directions, to allow the subscriber station to synchronize to the base station.

Also, the initial alerting of the base station about the presence of a subscriber station is problematic. As the subscriber station cannot decode the downlink map, an alerting slot with fixed location with respect to the frame preamble is foreseen. In this slot, the subscriber station can, possibly multiple times, send an alerting message.

Because of these complicated mechanisms, which have not been verified for their robustness, we believe that the greatest precautions should be taken in using AASs for range extension.

Mesh networking is an optional mode in the IEEE 802.16 standard, where a subscriber station can be reached from a base station in multiple hops (see *Figure 4.8*). As a consequence, mesh networking allows extending the range of a cell at the cost of its capacity. Indeed, if N_{hops} hops are needed to communicate from a subscriber station to a base station, and these N_{hops} communications interfere with each other, then the capacity is reduced with a factor N_{hops}. On the other hand, mesh networking is believed to reduce the necessary infrastructure investment and to simplify the network deployment.

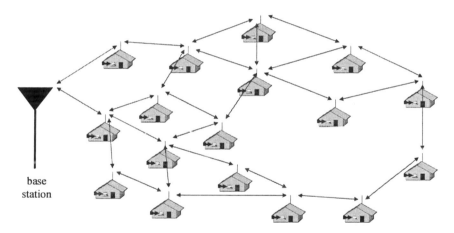

Figure 4.8. A mesh network extends the range of a cell at the cost of its capacity.

For the mesh networking, the IEEE 802.16 standard defines neighbourhoods, as shown in *Figure 4.9*. All stations that can be reached within one hop from a station form the neighbourhood of that station. All stations that can be reached in two hops are called the extended neighbourhood. An extensive set of messages is defined in the standard for the subscriber stations to signal the link and node identifiers of their neighbourhoods to each other. A station that has a backhaul connection to a fixed network is called a mesh base station.

The scheduling in a meshed network should be such that there are no collisions between communications in the extended neighbourhood. This scheduling can be performed in a distributed way in each subscriber station, or, centrally, in the mesh base station. Although the scheduling algorithm itself is not defined in the standard, the messages to communicate the schedule between the subscriber stations are provided.

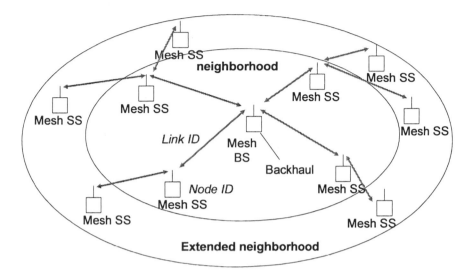

Figure 4.9. Mesh networks use two types of neighbourhoods.

For operation in the license-exempt frequency bands, the IEEE 802.16 standard imposes the use of Dynamic Frequency Selection (DFS). With DFS, each base station and subscriber station checks at start-up as well as periodically during operation for the presence of primary users in the frequency band. If primary users are found, the base station or subscriber station stops transmitting, and tries to switch to a different frequency band.

Next to the MAC function, the data link layer also contains a security sublayer. This sublayer provides encryption for the data packets that are transmitted on the air interface. As illustrated by the history of the Wired Equivalent Privacy (WEP) method that was originally used in the IEEE 802.11 WLAN standard, an encryption scheme with fixed keys can be easily broken [10]. Therefore, the IEEE 802.16 standard uses dynamic keys that are exchanged by means of a privacy key management protocol.

4.5 MAC EFFICIENCY

The overhead of the MAC signaling makes that the effective data rate for the users in an IEEE 802.16 cell is lower than the physical communication rate. This is expressed as the MAC efficiency, and, traditionally, authors approximate it by a constant value of 80% [11]. However, in practice, the

MAC efficiency is function of the physical communication mode, the packet size, the frame length, the number of up- and downlink users in a frame, and many more parameters of the MAC protocol. In this section, we therefore aim at obtaining a more accurate estimate of the MAC efficiency and at gaining some insight in its dependency on several key parameters.

We focus our investigation on a point-to-multipoint cell that employs an OFDM-based PHY with TDD (see Chapter 5 for more details on the PHY parameters). Furthermore, we make the following system assumptions:

- No payload header suppression is applied.
- ARQ is not used. As a consequence, we also do not need extended packing and fragmentation headers.
- Advanced antenna systems and space time coding are not used.
- We do not apply the optional CRC code.

The MAC efficiency, i.e. E_{MAC}, is analyzed analytically, where abstraction is made of possible packet errors. As a consequence, the results are slightly optimistic, as a zero Packet Error Rate (PER) is unrealistic. Every packet error will cause a retransmission at the application layer, and, hence, reduce the effective data rate. For small values of the PER, a good approximation of the resulting MAC efficiency, $E_{MAC,PER}$, as a function of the MAC efficiency when the PER equals 0, $E_{MAC,0}$, can be calculated as:

$$E_{MAC,PER} = E_{MAC,0} \cdot (1 - PER) \qquad (4.1)$$

The MAC efficiency is analyzed for various numbers of active users N_u. For this analysis, the MAC frame is constructed as follows. The downlink frame consists of a single burst, where the communications of the multiple subscriber stations are concatenated. This approach minimizes the length of the downlink map. For the uplink, we use a separate burst for every active subscriber station. The downlink and uplink maps are transmitted in the most robust communication mode, that is, BPSK with rate ½ coding.

The uplink subscriber stations request bandwidth by piggy backing a bandwidth request message on their uplink communication. Furthermore, to allow for inactive subscriber stations to request uplink bandwidth, we reserve in every frame a single OFDM symbol as bandwidth request slot.

In the analysis, we neglect the uplink and downlink channel description messages, as well as the initial ranging time slots. As these messages and time slots are normally not foreseen in every frame, they will only have a minor impact on the MAC efficiency.

Table 4.2 summarizes the values that were used for various parameters. For all other parameters, the default values of the standard were applied.

Table 4.2. Parameters for MAC efficiency calculation.

Parameter	Symbol	Value	Unit
bandwidth	BW	3.5	MHz
cyclic prefix length	N_{cp}	8	Samples (= OFDM symbol/256)
frame duration	t_f	10	ms
packet size	L_p	1500	Bytes
BW request slots	N_{BWreq}	1	OFDM symbol

The analysis approach consists of two steps. In a first step, the number of OFDM symbols per frame that are available for data transmission, N_{OFDM}, is estimated. This is achieved by means of the following formula:

$$N_{OFDM} = floor\left(\frac{T_f}{T_{OFDM}}\right) - N_{overheadsymbols}, \qquad (4.2)$$

where T_{OFDM} is the duration of one OFDM symbol (in ms), which is function of the bandwidth and the cyclic prefix length. The number of overhead OFDM symbols, $N_{overheadsymbols}$, grows with N_u and N_{BWreq}. We aim at distributing N_{OFDM} equally between up- and downlink, under the condition that all uplink users have an identical number of OFDM symbols.

In the second step, the effective number of data bits, N_d, that are transmitted in the uplink and downlink OFDM symbols, are estimated. This calculation is based on the following expression:

$$N_{data} = R_{mode} \cdot T_{OFDM} \cdot N_{OFDM} - N_{overheadbits} \qquad (4.3)$$

The variable $N_{overheadbits}$ equals the number of bits that are used for the MAC headers, packing headers, the piggy-backed bandwidth requests, and the termination bytes for the Forward Error Correction (FEC) coding. The physical communication rate for the various communication modes is represented by R_{mode}. For the assumptions of *Table 4.2*, the values of R_{mode} are summarized in the following table. The details of the modulation and coding schemes are explained in Chapter 5.

Table 4.3. Physical communication rates for the various communication modes.

Communication mode	1	2	3	4	5	6	7
Modulation	BPSK	QPSK	QPSK	16QAM	16QAM	64QAM	64QAM
Coding rate	1/2	1/2	3/4	1/2	3/4	2/3	3/4
R_{mode} (Mbps)	1,45	2,91	4,36	5,82	8,73	11,64	13,09

Once the number of effective data bits per frame is known, the MAC efficiency for a PER of zero, is calculated in a straight-forward way, with the help of following equation:

$$E_{MAC,0} = \frac{N_d}{T_f \cdot R_{mode}}$$ (4.4)

This MAC efficiency is plotted in *Figure 4.10*, as a function of the number of users and the communication modes. In contrast with the classical assumption of a constant MAC efficiency, we observe a large variation, ranging between 50% and 95%.

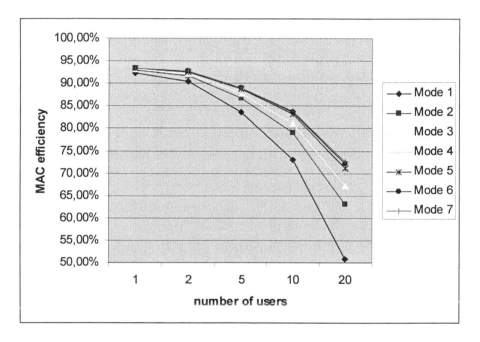

Figure 4.10. Efficiency of the 802.16 MAC in function of the physical communication mode and number of users.

From *Figure 4.10*, we can also conclude that the MAC efficiency degrades with the number of active users per frame. This behavior is explained by the increase in overhead with a growing number of users. The MAC efficiency is also lower for modulation schemes with smaller constellation sizes. This is caused by the growing importance of $n_{overheadbits}$, when the number of bits per OFDM symbol decreases.

The obtained results are well in line with corresponding results in [12]. In this paper, Hoymeyer et al. present the results of performance simulations of the IEEE 802.16a MAC layer. They analyze a system with a single user for the various communication modes and obtain a MAC efficiency of approximately 90%, which is independent of the physical communication mode. For a single user, we observe a similar behavior in the above graph, although with a slightly higher efficiency. The difference can be explained by the fact that, in contrast to our analysis, they assume the presence of a CRC code, as well as an initial ranging slot.

4.6 SUMMARY

In this chapter, we have introduced the IEEE 802.16 data link layer for broadband fixed wireless access, which comprises both the Medium Access Control (MAC) and the security sublayers. The MAC sublayer ensures that the communication resources are shared in a fair way between the different subscriber stations, while taking into account their various QoS requirements. The security sublayer defines the algorithms for encryption of the transmitted data packets, as well as the associated protocols for the exchange of the encryption keys.

We have pointed out that the service-specific convergence sublayers, which form the interface between the higher layer protocols and the data link layer, play a crucial role in the provisioning of QoS features. The main purpose of a convergence sublayer is to compare each incoming SDU with a set of classifiers that subsequently map the SDU on one of the open connections. Depending on the selected connection, part of the payload header can be suppressed, resulting in a higher effective data rate. To guarantee QoS, each connection is associated with a service flow, which is a description of the QoS parameters of a connection.

The MAC sublayer is very versatile, and supports a large number of optional features, including Automatic Retransmission reQuest (ARQ), support for Adaptive Antenna Systems (AASs), and a Mesh communication mode. Its centralized scheduling approach results in an effective use of the available physical communication rate. With less than 10 active users per MAC frame and QPSK (or higher-order) modulation, the MAC efficiency exceeds 80%.

4.7 REFERENCES

[1] N. Briscoe, "Understanding the OSI 7-Layer Model", PC Network Advisor, Issue 120, pp. 13-14, July 2000.

[2] C. Eklund, R. B. Marks, K. L. Stanwood, S. Wang, "IEEE Standard 802.16: A Technical Overview of the WirelessMAN™ Air Interface for Broadband Wireless Access", IEEE Communications Magazine, Vol. 40, No. 6, pp. 98-107, June 2002.

[3] SCTE DSS 00-05 [DOCSIS SP-RFIv1.1-I05-000714], Radio Frequency Interface 1.1 Specification, July 2000.

[4] IEEE Standard for Local and Metropolitan Area Networks, Part 16: Air Interface for Fixed Broadband Wireless Access Systems, IEEE 802.16-2001, 2001.

[5]   IEEE Standard for Local and Metropolitan Area Networks, Part 16: Air Interface for Fixed Broadband Wireless Access Systems, Amendment 2: Medium Access Control Modifications and Additional Physical Layer Specifications for 2-11 GHz, IEEE 802.16a-2003, 2003.

[6]   IEEE Standard for Local and metropolitan area networks, Part 16: Air Interface for Fixed Broadband Wireless Access Systems, Amendment 1: Detailed System Profiles for 10-66 GHz, IEEE 802.16c-2002, 2002.

[7]   IEEE Standard for Local and Metropolitan Area Networks Part 16: Air Interface for Fixed Broadband Wireless Access Systems, IEEE 802.16-2004 (Revision of IEEE 802.16-2001), 2004.

[8]   W. Jeong, M. Kavehrad, "Cochannel Interference Reduction in Dynamic-TDD Fixed Wireless Applications, Using Time Slot Allocation Algorithms" IEEE Transactions on Communications, Vol. 50, No. 10, pp. 1627 – 1636, October 2002.

[9]   A. Bourdoux, B. Côme, "Non-reciprocal Transceivers in OFDM/SDMA Systems: Impact and Mitigation", In proceedings of RAWCON 2003, Boston, pp. 183-186, August 2003.

[10] A. Stubblefield, J. Ioannidis, A. D. Rubin, "Using the Fluhrer, Mantin, and Shamir Attack to Break WEP", Proc. of ISOC Symposium on Network and Distributed System Security, San Diego, California, February, 2002.

[11] D. Kostas, M. Sellars, B. Freeman, G. Athanasiadou, "PHY Proposal for IEEE 802.16.3", contribution IEEE 802.16.3p-00/28, November 2000.

[12] C. Hoymann, M. Püttner, I. Forkel, "Initial Performance Evaluation and Analysis of the Global OFDM Metropolitan Area Network Standard IEEE 802.16a / ETSI HiperMAN", Proc. of the Fifth European Wireless Conference, Barcelona, Spain, February 24-27, 2004.

Chapter 5

Different PHYs Serve Different Needs
The BFWA Physical Layers

Frederik Petré

5.1 INTRODUCTION

Next to the data link layer, which was discussed in Chapter 4, the physical layer (PHY), that is, layer 1 in the OSI protocol layer stack, is also an essential ingredient of any BFWA solution. Its main purpose is to enable reliable data communication over the BFWA propagation channel, while coping with its main impairments, such as path loss, delay dispersion, time variance, and co-channel interference.

The IEEE 802.16 standard comprises several PHYs: one for line-of-sight communications between 10 and 66 GHz, and three for non-line-of-sight communications between 2 and 11 GHz [1]. The three non-compatible sub-11 GHz PHYs are the result of a compromise in the standardisation committee, where several contributors aimed at protecting their legacy solutions. In the European HiperMAN committee, there was a much stronger focus on interoperability of devices and, hence, only a single PHY, namely OFDM, was selected [2]. As a consequence, we expect that the OFDM PHY will also become dominant on the market for sub-11 GHz BFWA. In this chapter, we will therefore concentrate on the OFDM PHY, after having briefly reviewed the other PHY communication schemes.

This chapter is organized as follows. Section 5.2 gives a brief overview of the PHY for operation in the frequency bands between 10 and 66 GHz. Section 5.3 briefly introduces the three PHYs for communication in the frequency bands below 11 GHz, while Section 5.4 develops the OFDM PHY in more detail. Section 5.5 discusses the main OFDM PHY extensions to also support mobile next to fixed subscriber stations. Finally, Section 5.6 summarizes the main findings of this chapter.

5.2 PHYSICAL LAYER FOR 10-66 GHZ BFWA

The 10 to 66 GHz frequency bands are characterized by a physical propagation environment, in which line-of-sight operation is required and multipath propagation is negligible, due to the short wavelength. Channel bandwidths of 20 or 25 MHz, and 28 MHz are typical for U.S. allocation and European allocation, respectively, offering raw data rates in excess of 120 Mbps.

The single-carrier modulation air interface for 10 to 66 GHz operation, which is designated as WirelessMAN-SC [1][3], is a common Quadrature Amplitude Modulation (QAM) scheme. The constellation for this modulation scheme can have 4, 16, or 64 points. As a consequence, one symbol contains 2, 4, or 6 bits, respectively. The modulation symbols are filtered by a root raised cosine (RRC) pulse shaping filter with a roll off factor α of 0.25.

Before being transmitted, the bits are first protected against communication errors. Four different Forward Error Correcting (FEC) coding schemes are defined in the standard: a Reed-Solomon (RS) block code, the concatenation of a convolutional code with a RS block code, the combination of a RS block code with parity check codes, and, finally, block turbo codes. Before being encoded, the information bits are first randomized to allow for spectral shaping, and to ensure a sufficient number of bit transitions for clock recovery.

The PHY for 10 – 66 GHz also supports adaptive burst profiling, in which the modulation format and the coding scheme, or any relevant transmission parameter in general, may be adapted individually to each subscriber station on a burst-by-burst basis. This would allow the MAC to use bandwidth efficient burst profiles under favourable channel conditions, and resort to more robust, albeit less efficient burst profiles in less favourable situations.

Because of the point-to-multipoint architecture, the base station basically transmits a Time-Division Multiplexed (TDM) signal, in which time slots are allocated serially to the different subscriber stations. Access in the uplink direction is based on a combination of Time-Division Multiple Access (TDMA) and Demand Assignment Multiple Access (DAMA). The former refers to the division of the uplink channel into a number of disjoint time slots. The latter refers to the flexibility in the assignment of time slots for various uses, including registration, contention, guard, or user traffic, under the control of the MAC at the base station. To allow for flexible spectrum usage, both Time-Division Duplex (TDD) and Frequency-Division Duplex (FDD) configurations are supported. For TDD, the up- and downlink use the same carrier frequency while transmitting in disjoint time slots. In FDD, the up- and downlink transmit simultaneously while using different carrier frequencies.

5.3 PHYSICAL LAYERS FOR 2-11 GHZ BFWA

The 2 to 11 GHz frequency bands are characterized by a physical propagation environment, in which non-line-of-sight operation is feasible, and multipath propagation may be significant, due to the longer wavelength. The channel bandwidth, which is variable to ensure global applicability of the standard, can be an integer multiple of 1.25 MHz, 1.5 MHz, and 1.75 MHz, with a maximum of 20 MHz, corresponding to raw data rates up to 75 Mbps.

For the sub-11 GHz communication, the IEEE 802.16a standard defines three different PHYs: a single-carrier PHY (SCa), which is different from the high-frequency single-carrier PHY, an Orthogonal Frequency Division Multiplexing (OFDM) scheme, which goes a step further than IEEE 802.11a/g, and an Orthogonal Frequency Division Multiple Access (OFDMA) scheme, which is inspired on the standard for the return channel of Digital Video Broadcasting (DVB) [1][4][5].

This section is organized as follows. The single-carrier PHY is briefly discussed in Subsection 5.3.1, whereas the OFDM and the OFDMA PHYs are introduced in Subsection 5.3.2 and Subsection 5.3.3, respectively.

5.3.1 WirelessMAN-SCa

The single-carrier air interface, which is designated WirelessMAN-SCa , is based on Quadrature Amplitude Modulation (QAM) with constellations of 2 to 256 points [1]. Before modulation, the bits are protected against transmission errors with a concatenation of Reed-Solomon block coding and Trellis-coded modulation. Block turbo codes and convolutional turbo codes are defined as alternative options. Transmission bursts consist of a preamble, the payload data, and periodically inserted pilot words. As will be explained in Chapter 6, the periodic insertion of identical pilot words will allow for efficient frequency domain egalization, although this is not mandatory.

5.3.2 WirelessMAN-OFDM

The Orthogonal Frequency Division Multiplexing (OFDM) air interface, which is designated WirelessMAN-OFDM, relies on an OFDM symbol with 256 subcarriers, which is one step further than IEEE 802.11a/g that uses only 64 subcarriers [1]. Out of these 256 subcarriers, 192 are used for effective data transmission, 8 are used for continuous pilot symbols, and the remaining 56 are nulled for a guard band. Depending on the expected channel delay spread, a cyclic prefix of length 8, 16, 32, or 64 samples is prepended to the OFDM symbol, in order to provide robustness against delay dispersive multipath propagation.

Apart from the mandatory features, the WirelessMAN-OFDM PHY also includes a number of optional features. On the one hand, the use of Alamouti's space-time block coding scheme with two transmit antennas as well as the deployment of adaptive antenna systems are optional to improve the link reliability and/or to increase the spectral efficiency. These features will be explained in great detail in Chapter 7, which provides a comprehensive overview on smart antenna systems. On the other hand, the optional support of uplink sub-channelization enables low-cost, possibly mobile, subscriber stations, as will be discussed in Subsection 5.5.1.

From an implementation complexity point of view, the WirelessMAN-OFDM PHY seems to be favourable compared to the WirelessMAN-OFDMA PHY, for reasons such as lower peak-to-average-power ratio, faster FFT calculation, and less stringent frequency synchronization and phase noise requirements. The WirelessMAN-OFDM PHY will be discussed in more detail in Section 5.4.

5.3.3 WirelessMAN-OFDMA

The Orthogonal Frequency Division Multiple Access (OFDMA) air interface, which is referred to as WirelessMAN-OFDMA, uses an OFDM symbol with 2048 subcarriers [1]. In contrast to the WirelessMAN-OFDM PHY, not all 2048 subcarriers are assigned to a single user, but they are divided over 32 users. The OFDM symbol is prepended with a cyclic prefix with a variable length between ¼ and 1/32 of the OFDM symbol length. The FEC coding scheme is based on convolutional coding or, optionally, on block or convolutional turbo codes.

The very long symbols make the OFDMA PHY ideally suited for situations with long delay spread. The OFDMA PHY also allows to balance power among users and to work with terminals of limited transmit power.

The main disadvantage of the OFDMA PHY is its implementation complexity. As an example, for a small channel bandwidth of 1,5 Mhz the inter-subcarrier spacing will become only 732 Hz, resulting in very stringent requirements for the residual frequency offset (which must be below 2% of the inter-subcarrier spacing, i.e., approximately 15 Hz) and the phase noise.

5.4 WIRELESSMAN-OFDM IN MORE DETAIL

Because we expect that the BFWA market will be dominated by the WirelessMAN-OFDM PHY, we will elaborate it further in this section, which is organized as follows. Subsection 5.4.1 reviews the basic principles of OFDM modulation. Subsection 5.4.2 provides the details of the different building blocks of the WirelessMAN-OFDM PHY transmitter model. Finally, Subsection 5.4.3 makes a realistic comparison between single-carrier modulation and OFDM.

5.4.1 OFDM basics

Orthogonal Frequency Division Multiplexing (OFDM) is a multi-carrier modulation technique that capitalizes on the Fast Fourier Transform (FFT) and the addition of transmit redundancy in the form of a cyclic prefix [6][7][8][9]. By relying on IFFT/FFT operations, it transmits the information symbols on multiple subcarriers. By inserting a cyclic prefix at the transmitter and discarding it again at the receiver, these subcarriers preserve their shape and orthogonality, even after propagation through a frequency-selective multipath channel. As such, OFDM divides the available physical bandwidth BW into N_c subcarriers or frequency bins. By choosing N_c sufficiently high, the information bandwidth per subcarrier is narrow compared to the coherence bandwidth of the channel, that is, $BW / N_c << B_{coh}$, such that each subcarrier experiences frequency-flat fading. The working principle of OFDM is illustrated in *Figure 5.1*.

In contrast with single-carrier modulation, which transmits the information symbols in a serial fashion and traditionally requires a large time-domain equalizer to mitigate ISI, OFDM only needs a low-complexity frequency-domain equalizer, consisting of a single-tap equalizer per subcarrier.

Because of its effective handling of broadband multipath propagation, OFDM has been adopted in the BFWA standard. In order to also exploit the inherent frequency diversity of the BFWA multipath channel, FEC coding is performed over the different subcarriers or frequency bins. An alternative strategy is to adapt the modulation format, and, hence, the number of bits, per subcarrier, depending on the experienced Signal-to-Noise Ratio (SNR) on each subcarrier [10]. This strategy is, for instance, followed in

Asymmetric Digital Subscriber Line (ADSL) technology [11], but has not yet been included in the WirelessMAN-OFDM scheme.

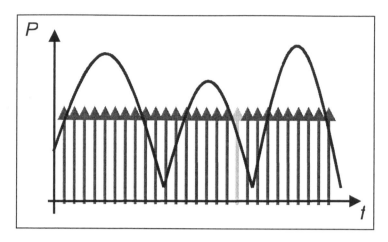

Figure 5.1. OFDM splits data over multiple subcarriers.

5.4.2 WirelessMAN-OFDM PHY transmitter model

The purpose of the PHY is twofold. In transmit mode, it should map the Protocol Data Units (PDUs) it receives from the data link layer into PHY bursts, which can be transmitted and received over the air interface. In receive mode, it should process the received data burst, in order to obtain reliable estimates of the transmitted PHY burst data and map the result into PDUs, which can subsequently be passed on to the data link layer.

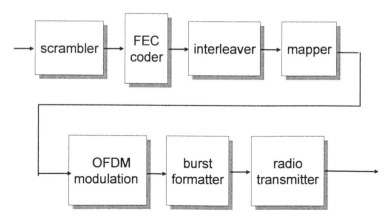

Figure 5.2. Overview of the WirelessMAN-OFDM PHY transmitter model.

As in most standards, the receive part of the WirelessMAN-OFDM PHY is only specified through overall quality requirements, in order to leave room for differentiation between competitors. Therefore, we focus on the transmit specification, which is illustrated in *Figure 5.2*. The standard defines an OFDM transmitter that consists of seven blocks: the data scrambler, the Forward Error Correction (FEC) coder, the data interleaver, the symbol mapper, the OFDM modulator, the burst formatter, and, finally, the radio transmitter. First, the incoming PDU train is transformed into a pseudo-random bitstream using a scrambling sequence. The resulting scrambled PDU train is then protected against transmission errors with the aid of a FEC code. Subsequently, the encoded bitstream is interleaved in order to spread error bursts at the receiver over the PHY packet. The bits are then mapped to complex modulation symbols on each subcarrier according to the modulation alphabet, after which the subcarriers are OFDM modulated using the IFFT. The resulting OFDM symbols are formatted into a PHY burst, which contains an additional preamble. Finally, this PHY burst is transmitted over the air by modulating it on the radio frequency carrier.

In the following subsections, we will discuss each of these blocks in more detail.

5.4.2.1 Data scrambler

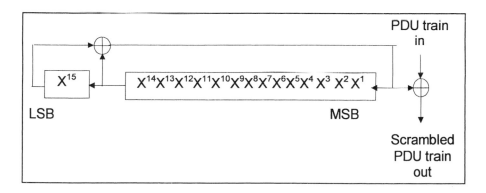

Figure 5.3. WirelessMAN-OFDM data scrambler.

As shown in *Figure 5.3*, the data scrambler contains a pseudo-random sequence generator. The incoming PDU train is exored, or scrambled, with this pseudo-random scrambling sequence. The result is an outgoing scrambled PDU train.

The random sequence generator is initialized at the start of each frame and each burst. The downlink initialization is based on the base station identifier, the downlink interval usage code, and the frame number.

Likewise, the uplink initialization is based on the base station identifier, the uplink interval usage code, and the frame number.

5.4.2.2 FEC coding parameters

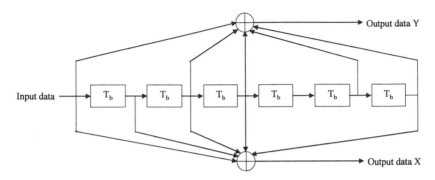

Figure 5.4. WirelessMAN-OFDM convolutional encoder.

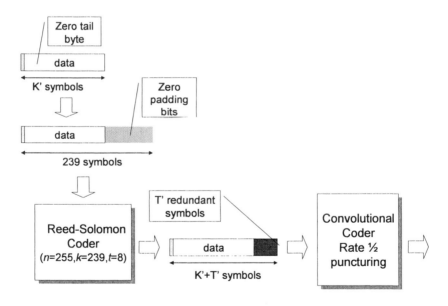

Figure 5.5. A zero-tail byte is used to re-initialize the convolutional coder.

The default FEC coding scheme is the concatenation of a Reed-Solomon (RS) block code and a convolutional code. The RS code works on blocks of 255 symbols, and can correct up to 8 errors. By puncturing the code, the number of redundant symbols is reduced, at the cost of a reduced error correcting capability. By shortening the code, it can work on shorter blocks of symbols. In the standard, the RS code is always shortened, such that it works on a block of symbols that corresponds exactly to the data for one OFDM symbol.

The second element of the concatenated coder consists of a standard rate ½ convolutional encoder, which is illustrated in *Figure 5.4*. To reduce the overhead, puncturing patterns are foreseen, to obtain codes with rates 2/3, ¾, and 5/6.

To simplify the coding, the convolutional coder is re-initialized with all zeros, at the end of each burst. This is achieved by adding a zero tail byte to the data, and place the redundant RS-bits before the data bits. As such, the redundant tail byte, which is at the end of the RS output, re-initiliazes the convolutional coder for the next burst. This principle is schematically shown in *Figure 5.5*.

Table 5.1. Transmission modes for WirelessMAN-OFDM.

Modulation	Uncod.block size (bytes)	Coded block size (bytes)	Overall coding rate	RS code rate	CC code rate
QPSK	24	48	½	{32,24,4}	2/3
QPSK	36	48	¾	{40,36,2}	5/6
QAM-16	48	96	½	{64,48,8}	2/3
QAM-16	72	96	¾	{80,72,4}	5/6
QAM-64	96	144	2/3	{108.96,6}	¾
QAM-64	108	144	¾	{120,108,6}	5/6

By combining different puncturing and truncating rates for the RS coder, and different code rates for the convolutional coder, overall coding rates of ½, 2/3, and ¾ are achieved. Combined with the three QAM modulation orders, the standard selects 6 transmission modes, with increasing throughput, but also more stringent SNR requirements. *Table 5.1* summarizes these standardized transmission modes.

In addition to the mandatory FEC coding scheme, the standard also describes two optional turbo coding schemes. In the block turbo code, the data is ordered in a square, and block codes are sequentially performed in the horizontal and vertical directions. In the convolutional turbo code, two convolutional codes are performed in parallel and the data is interleaved before going to the second convolutional encoder. *Figure 5.6* compares the bit error rate (BER) versus SNR performance of the three coding schemes over the SUI-3 channel, when combined with QAM-16 modulation. If we compare the performance of both optional turbo coding schemes, we observe that the convolutional turbo code outperforms the block turbo code for a coding rate of 2/3. The gain compared to the mandatory scheme is about 4 dB. Remark that block turbo codes are more advantageous for very high coding rates. However, these are not supported in the standard.

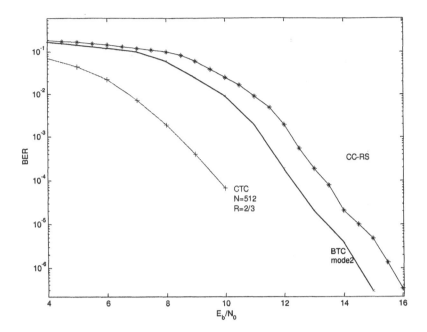

Figure 5.6. Comparison of three different FEC coding schemes: concatenation of Reed-Solomon and Convolutional Coding (CC-RS), Block Turbo Codes (BTC) and Convolutional Turbo Codes (CTC).

5.4.2.3 Data interleaver

After FEC coding, the data interleaver applies a two step permutation on the data that is transmitted in a single OFDM symbol. The first step ensures that adjacent coded bits are mapped onto non-adjacent subcarriers. By spreading adjacent bits sufficiently apart, that is, further than the coherence bandwidth, B_{coh}, they will experience independent fading, resulting in a reduced probability of burst error. Let us denote by k the index of the coded bit before the first permutation, by m and j the index after the first and the second permutation, respectively. The first permutation is then defined by:

$$m = \frac{N_{CBPS}}{12}(k \bmod 12) + \left\lfloor \frac{k}{12} \right\rfloor, \tag{5.1}$$

where N_{CBPS} represents the number of coded bits per OFDM symbol. Hence, the OFDM symbol is divided into 12 groups of subcarriers, and adjacent coded bits are mapped onto different groups.

The second permutation ensures that adjacent coded bits are mapped alternately onto less and more significant bits of the constellation. The motivation for this alternation is justified by the fact that the Least Significant Bits (LSBs) have a higher probability of error than the Most Significant Bits (MSBs), when a symbol is faulty. In this way, long runs of low reliability bits are avoided. The second permutation can be defined as:

$$j = s \left\lfloor \frac{m}{s} \right\rfloor + \left(m + N_{CBPS} - \left\lfloor \frac{12m}{N_{CBPS}} \right\rfloor \right) \bmod s , \qquad (5.2)$$

where $s = N_{CBPS}/2$.

5.4.2.4 Symbol mapping

After FEC coding and interleaving, the coded bits are mapped onto Quadrature Amplitude Modulation (QAM) symbols. To minimize the probability of bit errors, the mapper uses Gray coded constellations. Four different constellations are supported: BPSK, QPSK, 16-QAM, and 64-QAM. The average power for each of these constellations is normalized. *Figure 5.7* shows as an example the Gray mapping for a 16-QAM constellation. The normalization factor for this mapping is $1/\sqrt{10}$.

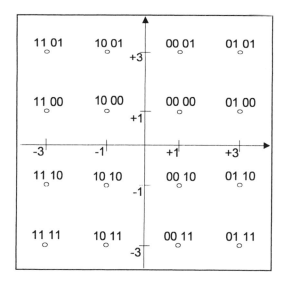

Figure 5.7. Gray mapping for 16-QAM constellation.

5.4.2.5 OFDM modulation

The QAM symbols are next put on the data subcarriers. With 256 subcarriers available, indexed from −128 to +127, the used subcarriers range from index −100 to index +100. The other subcarriers are used as guard band to reduce the interference to neighbouring bands. From the used subcarriers, 8 are reserved to transmit known pilot data. These pilot subcarriers can be used in the receiver to compensate small frequency offsets and slow variations of the propagation channel. Moreover, the DC subcarrier with index 0 is not used, in order to avoid overloading of the analog front-end. As a consequence, 192 subcarriers out of the 256 effectively transmit data.

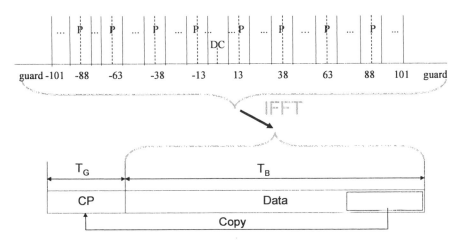

Figure 5.8. Subcarrier frequency assignment and guard interval in WirelessMAN-OFDM.

As illustrated in *Figure 5.8*, the resulting frequency domain representation is transformed into a time domain signal with 256 time samples by performing an Inverse Fast Fourier Transform (IFFT). Next, a cyclic prefix is prepended to this OFDM symbol. The length of the cyclic prefix should be long enough to encompass the delay spread of the propagation channel. On the other hand, it should be minimized, as it constitutes an overhead with respect to power and data rate. Therefore, the standard offers flexibility and supports four lengths of the cyclic prefix: 64, 32, 16, or 8 time samples.

In summary, the basic parameters for the WirelessMAN-OFDM PHY are listed in *Table 5.2*. The number of subcarriers is 256, of which 56 are guard subcarriers and 200 are effectively used. From the latter, 8 subcarriers transmit known pilot data. The 192 effective data subcarriers can be divided in 16 subchannels that each consist of 4 groups of 3 subcarriers. For the cyclic prefix length, some flexibility is provided, ranging from ¼ to 1/32 of an OFDM symbol. The ratio between the sampling rate and the bandwidth is function of the bandwidth. For instance, for bandwidths that are a multiple of 1.75 MHz, the ratio is 8/7.

Table 5.2. OFDM parameters for WirelessMAN-OFDM.

Parameter	Value
Number of subcarriers	256
Number of used subcarriers	200
F_s/BW	Depens on BW. 8/7 for multiples of 1.75 MHz
Cyclic prefix length	¼, 1/8, 1/16, 1/32
Guard subcarriers	{-128,...,-101},0,{101,...,127}
Pilot subcarriers	{-88,-63,-38,-13,13,38,63,88}
Number of subchannels	16 (each one a group of 4 x 3 subcarriers)

5.4.2.6 Burst formatting

Transmission bursts consist of a preamble and a payload section. The preamble structure differs according to the burst type, whereas the payload in all cases consists of the OFDM-modulated data symbols. The preamble is used for initial detection of the signal, the signal strength estimation, the estimation of the carrier frequency offset, and the estimation of the propagation channel coefficients. The preambles have been designed such that they feature a low Peak-to-Average Power Ratio (PAPR). By doing so, it is guaranteed that the non-linearity of the power amplifier will not affect the performance of the Automatic Gain Control (AGC) mechanism.

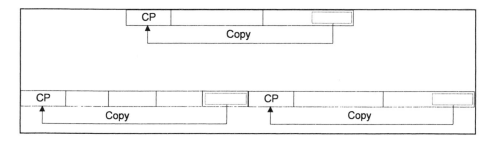

Figure 5.9. Short preamble (1 OFDM symbol) versus long preamble (2 OFDM symbols).

Two kinds of preamble structures are defined, which are illustrated in *Figure 5.9*. Short preambles consist of a single OFDM symbol with two identical halves. They are intended for downlink bursts, where it can be assumed that the symbol timing, frequency, and gain are already synchronized between transmitter and receiver. In all other cases, the long preambles that comprise two OFDM symbols are used. Here, the second OFDM symbol of the long preamble is identical to the short preamble, and the first OFDM symbol consists of four identical quarter symbols.

5.4.2.7 Radio transmitter

The task of the radio transmitter is to upconvert the signal to the carrier frequency, amplify it with a controllable gain, and put it on the air. Because of the wide variety of available frequency bands and regulatory provisioned bandwidths, the frequency plan is rather flexible. Basically, it is based on multiples of 250 kHz for the bandwidth and spacing between the carrier frequencies.

For the transmitter spectral mask, the standard simply refers to the local regulatory requirements. However, it specifies the in-band flatness: the average energy of the constellations shall not deviate more than ± 2dB for carriers between -50 and 50, and +2 dB to − 4dB for the other carriers, from the average energy of all carriers.

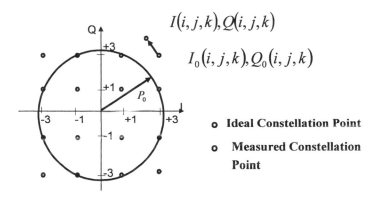

Figure 5.10. Definition of constellation points for Error Vector Magnitude calculation.

The most interesting radio requirement is the transmitter constellation error. It defines an upper bound for the errors that are introduced by the transmitter radio, as expressed by following formula:

$$ERROR_{RMS} = \dfrac{\displaystyle\sum_{i=1}^{N_p}\sqrt{\dfrac{\displaystyle\sum_{j=1}^{L_p}\sum_{\substack{k=-100\\k\neq0}}^{100}\left([I(i,j,k)-I_0(i,j,k)]^2+[Q(i,j,k)-Q_0(i,j,k)]^2\right)}{200\cdot L_p\cdot P_0}}}{N_p}, \quad (5.3)$$

with the number of packets N_p and the number of OFDM symbols in a packet L_p. The other parameters of this formula are clarified in *Figure 5.10*.

5.4.2.8 Data rates

Different bandwidths, cyclic prefix lengths, modulation orders and coding rates, result in different data rates. *Table 5.3* gives an overview of various options, ranging from 2.68 Mbps in a 3.5 MHz bandwidth to 53.46 Mbps in a 14 MHz bandwidth. Remark that these data rate figures express the raw data rate on the channel. To obtain the net data rate, these figures should be multiplied by the MAC efficiency, which is typically 80 %.

Table 5.3. Typical data rates for WirelessMAN-OFDM.

BW (MHz)	Cyclic prefix length	Bitrate QPSK ¾	Bitrate 16-QAM ¾	Bitrate 64-QAM ¾
3.5	1/32	4.46	8.91	13.37
3.5	¼	2.68	7.35	11.03
7	1/32	8.91	17.82	26.73
7	¼	7.35	14.70	22.05
14	1/32	17.82	35.64	53.46
14	¼	14.70	29.40	44.10

5.4.3 SCa versus OFDM

As the HiperMAN standard aimed to reduce the number of PHYs, an extensive comparison was made between OFDM and single-carrier modulation [2]. The main arguments against OFDM were the high ratio between the peak power and the average power. As can be seen in *Table 5.4*, some claimed a significant advantage for single-carrier, especially for lower-order modulations [12][13]. Due to this high PAPR, the non-linearity of the power amplifier would result in out-of-band emissions.

Table 5.4. Simulated PAPR of SCa versus OFDM.

Modulation order	SCa	OFDM
QPSK	7.5 dB	12.0 dB
16-QAM	9.4 dB	12.0 dB
64-QAM	10.5 dB	12.3 dB

However, both claims did not take clipping into account, which is generally applied at the transmitter. Including this clipping, the PAPR reduces to 8 dB, and the out-of-band spectrum is eliminated. Hence, none of the two communication schemes had a decisive argument. However, a considerable group of companies supported the OFDM PHY, and closely cooperated to improve it. The single-carrier PHY, on the other hand, was less well elaborated, and was rather driven by the desire to incorporate legacy solutions. After several meetings, the OFDM PHY was finally selected for the HIPERMAN standard. The IEEE 802.16 standard includes both incompatible options.

5.5 EXTENSIONS TOWARDS MOBILITY

At the time of writing of this book, a significant effort was going on in the IEEE 802.16 Broadband Wireless Access Working Group to revise the standard to also support mobile next to fixed subscriber stations. The resulting revised standard should be interpreted as a draft amendment to the baseline standard [14]. Specific to the WirelessMAN-OFDM PHY, two mobility extensions have already been included in the IEEE 802.16-2004 standard, to enable a smooth transition toward truly mobile subscriber stations. These are uplink subchannelization, on the one hand, and signal support for estimation of the downlink mobile channel, on the other hand.

This section is organized as follows. Subsection 5.5.1 introduces uplink subchannelization as a way to enable low-cost subscriber stations, through the use of power amplifiers with lower transmission power. Subsection 5.5.2 describes the revised transmitted signal structure in terms of midambles and/or pilots, to facilitate the estimation of the downlink mobile channel.

5.5.1 Uplink subchannelization

To achieve a major take up of 802.16 technology, low-cost terminals are essential. One major aspect of cost reduction is the use of power amplifiers with lower transmission power. This would severely limit the transmission range for a cell, unless the available power could be concentrated on less subcarriers. This consideration was one of the reasons for the standardization committee to include the option that the uplink OFDM symbol can be divided in up to 16 subchannels. Each subchannel can be used by a different terminal, effectively realizing an OFDMA scheme. To support low power customer premises equipment (CPE), extra provisions had to be made in the initial ranging and bandwidth requesting mechanisms. Besides supporting low power terminals, subchannelization also reduces the granularity of the bursts that are allocated to different users. As a consequence, it increases the efficiency of the MAC protocol, especially for frequent small packages, as for instance encountered in voice communications. Extensive simulations have been performed with respect to the impact of subchannelization to the implementation complexity. For instance, the phase noise specifications were extensively simulated, but no major complexity increase has been found.

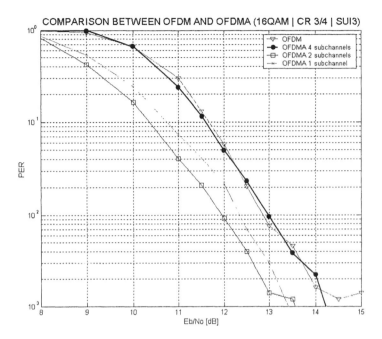

Figure 5.11. Comparison between OFDM and OFDMA/subchannelization with 1, 2, and 4 subchannels.

Because of the reduced packet sizes for subchannelization, the block based RS coding is not very effective. Therefore, when subchannelization is used, only convolutional coding is applied. *Figure 5.11* shows the Packet Error Rate (PER) of this error coding scheme when the OFDM symbol is divided in four subchannels, and the terminal uses 1, 2, or 4 of them for transmission. The packet length is 40 bytes, and the OFDMA scheme employs lumped subcarriers. As can be concluded from *Figure 5.11*, the PER of the subchannels even outperforms the standard OFDM PHY.

As illustrated in *Figure 5.12*, the selection of the subcarriers for the subchannels can be done in two different ways. The lumped approach selects 12 neighboring subcarriers, whereas the clustered approach combines smaller groups of subcarriers. The latter has a performance benefit because of the increased frequency diversity. Therefore, the standard opted for this clustered approach with four clusters of three subcarriers for every subchannel. Remark that with subchannelization, not every of the 16 subchannels can have a pilot tone. Therefore, a decision feedback scheme will have to be used for updating the channel estimation.

Finally, it should be noted that the IEEE 802.16-2004 standard also foresees subchannelization in the downlink, in order for communication resources to be allocated with a finer granularity [1].

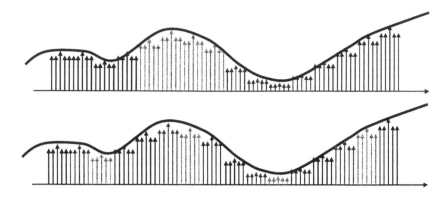

Figure 5.12. Lumped versus clustered subcarrier allocation.

5.5.2 Support for downlink mobile channel estimation

Two possible modifications can be made to the downlink signal structure of the WirelessMAN-OFDM PHY, to tackle the issue of estimating the time-varying channel at the mobile subscriber station, namely, the midamble versus the hopping-pilots scheme [15].

The midamble scheme inserts an additional midamble symbol, once every L regular symbols. Furthermore, it is assumed that these midambles have the same basic structure as the regular short preambles, that is, a length-256 OFDM symbol with only the even subcarriers occupied and a 3 dB boost in overall symbol energy. A clear tradeoff arises between the transmission overhead versus the estimation performance. Indeed, as L grows larger, the transmission overhead ($\sim 1/L$) is reduced, while the estimation performance is degraded. The midamble scheme can be used in two different operating modes. In the first operating mode, channel estimation is performed for every symbol, by relying on the two midambles adjacent to the current symbol. In the second operating mode, channel estimation is performed only when a midamble is received, and based on that midamble only. Furthermore, the channel estimate is held constant untill the next midamble is received. The former operating mode entails a latency of up to L symbols, whereas the latter incurs no additional latency.

The hopping pilots scheme capitalizes on the already existing pilots by changing, or hopping, their location from one symbol to the next. Recall that the WirelessMAN-OFDM PHY uses length-256 OFDM symbols with 200 active subcarriers, of which 192 are effective data subcarriers, and 8 are pilot subcarriers at fixed locations. However, in the hopping-pilots scheme, the location of these 8 pilot subcarriers is varied cyclically from one symbol to the next, with a period of 8 symbols. Consequently, no additional overhead is needed to support this scheme.

Extensive performance simulations of the different schemes have revealed the following three major conclusions [15]. First, when used in the second operating mode, the midamble scheme fails completely at the vehicular speeds examined. Second, when used in the first operating mode, the midamble scheme incurs a significant overhead to achieve similar performance as the hopping-pilots scheme. For instance, at a speed of 150 km/h, a midamble spacing of 6 symbols is required, which translates into a 17 % overhead. Finally, the hopping-pilots scheme outperforms the midamble scheme, without incurring any additional overhead. Nevertheless, only the midamble scheme has been included in the IEEE 802.16-2004 standard.

5.6 SUMMARY

In this chapter, we have provided a bird's eye view of the different physical layers for broadband fixed wireless access, as described in the IEEE 802.16 standard. This standard comprises several PHYs, each of them satisfying different needs in terms of the operating environment. On the one hand, the WirelessMAN-SC PHY is suited for communication in the 10 to 66 GHz frequency bands, which is characterized by line-of-sight operation and negligible multipath propagation.

On the other hand, three non-compatible PHYs have been developed for communication in the 2 to 11 GHz frequency bands, which is characterized by non-line-of-sight operation and significant multipath propagation. The WirelessMAN-SCa PHY, which differs from the high-frequency WirelessMAN-SC PHY, is mainly intended for near line-of-sight operations. The two OFDM-based PHYs, namely WirelessMAN-OFDM and WirelessMAN-OFDMA, mainly target non-line-of-sight operation including severe multipath propagation, with the latter being more suitable for environments with very long delay spread. Nevertheless, for reasons of interoperability between devices, it is widely believed that the WirelessMAN-OFDM PHY will become dominant in the BFWA market.

Currently, a significant effort is going on in the IEEE 802.16 Broadband Wireless Access Working Group to provide enhancements to the baseline standard, to also support mobile (besides fixed) subscriber stations moving at vehicular speeds. Two important mobility enhancements to the WirelessMAN-OFDM PHY include uplink subchannelization and signal support for downlink mobile channel estimation. Uplink subchannelization, which divides the uplink OFDM symbol in up to 16 subchannels, enables low-cost terminals, through the use of power amplifiers with lower maximum output power. The midamble scheme, which inserts an additional midamble symbol, greatly facilitates the estimation process of the time-varying downlink channel, at the expense of a reduced effective data rate.

5.7 REFERENCES

[1]   IEEE 802.16-2004, "IEEE Standard for Local and Metropolitan Area Networks – Part 16: Air Interface for Fixed Broadband Wireless Access Systems", Revision of IEEE 802.16-2001, December 2004.

[2]   ETSI TS 102 177, "Broadband Radio Access Networks (BRAN) – HiperMAN – Physical layer", RTS/BRAN-0040001r3, Version 1.2.2, November 2005.

[3]   C. Eklund, R. B. Marks, K. L. Stanwood, S. Wang, "IEEE Standard 802.16: A Technical Overview of the WirelessMAN Air Interface for Broadband Wireless Access", IEEE Communications Magazine, Vol. 40, No. 6, pp. 98-107, June 2002.

[4]   I. Koffman, V. Roman, "Broadband Wireless Access Solutions Based on OFDM Access in IEEE 802.16", IEEE Communications Magazine, Vol. 40, No. 4, pp. 96-103, April 2002.

[5]   A. Gosh, D. R. Wolter, J. G. Andrews, R. Chen, "Broadband Wireless Access with WiMax/802.16: Current Performance Benchmarks and Future Potential", IEEE Communications Magazine, Vol. 43, No. 2, pp. 129-136, February 2005.

[6]   J.A.C. Bingham, "Multicarrier Modulation for Data Transmission: An Idea Whose Time Has Come", IEEE Communications Magazine, Vol. 28, No. 5, pp. 5-14, May 1990.

[7]   A.R.S. Bahai, B.R. Saltzberg, "Multi-Carrier Digital Communications: Theory and Applications of OFDM", Kluwer Academic Publishers, 1999.

[8]   M. Engels (editor), "Wireless OFDM Systems: How to Make Them Work?" Kluwer Academic Publishers, 2002.

[9]   Z. Wang, G.B. Giannakis, "Wireless Multicarrier Communications: Where Fourier Meets Shannon", IEEE Signal Processing Magazine, Vol. 17, No. 3, pp. 29-48, May 2000.

[10]   P.S. Chow, J.M. Cioffi, J.A.C. Bingham, "A Practical Discrete Multitone Transceiver Loading Algorithm for Data Transmission over Spectrally Shaped Channels", IEEE Transactions on Communications, Vol. 73, No. 2/3/4, pp. 773-775, February/March/April 1995.

[11]   P. Reusens, D. Van Bruyssel, J. Sevenhans, S. Van Den Bergh, B. Van Nimmen, P. Spruyt, "A Practical ADSL Technology Following a Decade of Effort", IEEE Communications Magazine, Vol. 39, No. 10, pp. 145-151, October 2001.

[12]   P. Struhsaker, K. Griffin, "Analysis of PHY Waveform Peak to Mean Ratio and Impact on RF Amplification", IEEE 802.16 Broadband Wireless Access Working Group, IEEE802.16.3c-01/46, March 2001.

[13]   J. Tubbax, B. Côme, L. Van der Perre, L. Deneire, M. Engels, "OFDM versus Single-Carrier with Cyclic Prefix: a System-Based Comparison", Proceedings of IEEE VTC-Fall, Vol. 2, pp. 1115-1119, October 2001.

[14]   IEEE P802.16e/D8-2005, "Draft Amendment to IEEE Standard for Local and Metropolitan Area Networks – Part 16: Air Interface for Fixed Broadband Wireless

Access Systems – Physical and Medium Access Control Layers for Combined Fixed and Mobile Operation in Licensed Bands", May 2005.

[15] IEEE 802.16 Broadband Wireless Access Working Group, "Comparison between the Midamble and the Hopping-Pilots Scheme for Estimation of the Downlink Mobil Channel", IEEE C802.16e-03/18, March 2003.

Chapter 6

How to Realise a Superior Modem?
Implementation Aspects of an IEEE 802.16 Transceiver

Marc Engels

6.1 TRANSCEIVER DESIGN PROCESS

Low cost standard-compliant terminals that have a good performance in a multipath propagation environment are an essential condition for the breakthrough of WiMax technology. The standard leaves many degrees of freedom to design such terminals. Indeed, only the transmission formats are defined by the standard, while several algorithms, e.g. for the receiver, have to be defined during the conception of the terminal. As a consequence, the terminal design process can be viewed as an optimization process, during which the cost and, in the case of mobile solutions, the power consumption are traded off against performance. For the physical layer (PHY) of a communication system, this performance is often expressed as an implementation loss (IL), or a bit error rate (BER) degradation, as illustrated in *Figure 6.1*. For both measures, an ideal reference BER versus SNR curve is compared with a real BER versus SNR curve that includes the specific implementation choices. The IL is defined as the difference in SNR, expressed in dB, between these two curves at a specific BER, which should correspond to the typical operating value. Similarly, the BER degradation is the difference in BER between these two curves at a specific SNR that corresponds to a typical operating condition. The IL and BER degradation are function of the propagation model as well as the PHY communication mode. By using any of these two performance measures, different implementation choices with varying impact on cost and power consumption can be compared.

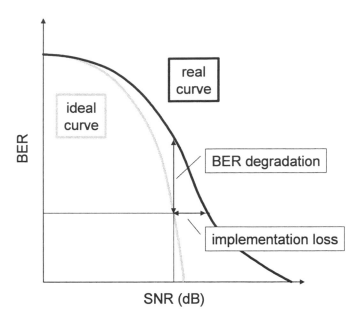

Figure 6.1. The implementation loss and BER degradation are often used as measures for the performance of a PHY implementation of a communication system.

Similarly, the performance of a data link layer is often visualized by a curve that expresses the average packet latency (in s) versus the throughput (in bits/s), as illustrated in *Figure 6.2.* Consequently, the effect on the communication performance of implementation choices at the data link layer can be quantified by means of the throughput, or latency, degradation. Again, an ideal reference curve is compared with a real curve that includes the specific implementation choices. The throughput degradation is defined as the difference in throughput between these two curves at a specific average latency that corresponds to the typical operating condition. The latency degradation is the difference in average packet latency between these two curves for the target throughput of the system. The latency and throughput degradations are function of a large number of parameters, including the packet sizes, the service mix, the number of users, etc. The measures are also function of the performance of the PHY. Therefore, this layer is often abstracted to a simplified model with random packet errors and a fixed packet error rate (PER).

As a consequence of the use of these performance measures, the design of an optimized terminal starts with a reference simulation of the standard communication scheme, both at the PHY and the data link layer. The performance of these simulations, based on realistic propagation conditions but with ideal receiver processing and data link algorithms, is used as a reference for the further optimization. The goal of the further design is to minimize the implementation cost and power consumption, given an implementation loss, BER degradation, latency degradation, and throughput degradation that can be tolerated.

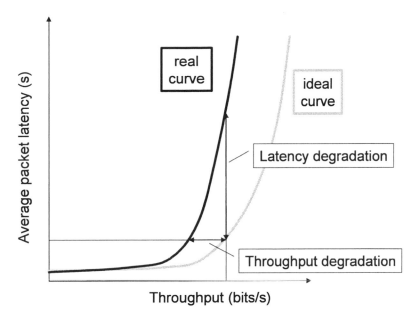

Figure 6.2. The latency or throughput degradation could be used to express the performance of a data link layer implementation of a communication system.

As shown in *Figure 6.3*, the design flow of the transceiver continues with the addition of the missing algorithms at the data link layer and the PHY, respectively. On the one hand, at the data link layer, this includes the design of an efficient search method for the classifiers in the service-specific convergence sublayer, a scheduling algorithm that guarantees QoS constraints [2], a mechanism to switch between the various PHY communication modes, and a dynamic frequency selection (DFS) approach. Each of these algorithms will have an important impact on the latency and throughput of the data link layer. Because of the multi-user aspects, the processing load for these algorithms is much higher at the base station compared to the subscriber station. As a consequence, asymmetric architectures will result that need more processing power in the base station.

On the other hand, for the PHY, two important receiver functionalities need to be defined: the channel estimation method and the synchronization strategy. Both these digital signal processing (DSP) algorithms have a significant impact on the implementation loss and the BER degradation. An even greater challenge in the transceiver design is the mixed signal optimization of the radio front-end specification and the corresponding digital compensation techniques. Indeed, many of the shortcomings of analog radio front-ends can be compensated by appropriate DSP in the digital domain. This results in superior transceiver performances, even with worse radio solutions, leading to a lower cost and power consumption. Compensation of the radio front-end impairments can be performed both at the transmitter and the receiver. In addition, techniques to control the dynamic range of a signal can be exploited at the transmitter to simplify the requirements towards the (analog) power amplifier.

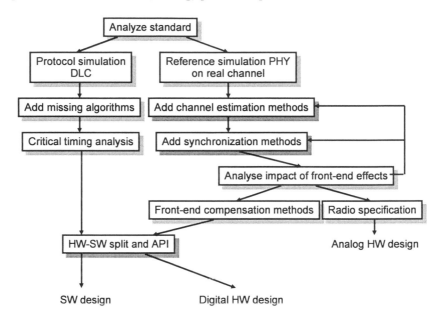

Figure 6.3. Design Flow for integrated digital communication systems.

Once all algorithms have been fixed, the designer will focus on the actual architectural design of the transceiver. As a preparation, the critical timings of the data link layer should be analyzed. In the WiMax standard, for instance, the delay between the reception of the uplink map and the first uplink transmission determines the critical path in receiving and transmitting a packet. For this particular parameter, the standard guarantees a minimum timing of 200 μs, 1 ms, or 10 OFDMA symbols, respectively, for the WirelessMAN-SCa, WirelessMAN-OFDM, and WirelessMAN-OFDMA PHY modes. Based on this timing analysis, the split between hardware (HW) and software (SW) can be decided upon, resulting in a digital HW architecture. From the radio specification, the analog HW architecture and the detailed properties of its building blocks are derived.

For an extensive discussion of this transceiver design process, we refer to [3]. In this chapter, we will restrict the discussion to a high-level overview with a particular focus on the most relevant implementation aspects of a IEEE 802.16 compliant transceiver. Section 6.2 explains the top level architectural choices for a WiMax transceiver. Next, Section 6.3 discusses the different baseband tradeoffs, with an emphasis on synchronization and channel estimation. This section also highlights the similarities between the implementations of the WirelessMAN-SCa and the WirelessMAN-OFDM PHY modes. Section 6.4 focuses on the radio front-ends, along with its analog impairments, and the associated digital compensation techniques. Finally, Section 6.5 concludes the chapter with a summary of the main observations.

6.2 TOP LEVEL ARCHITECTURE

Because of the similarity between the IEEE 802.16 and DOCSIS cable modem [4] standards, there will also be a large resemblance between the top-level architecture of both equipments. In [5], an extensive overview is given of the MAC functionality of a cable modem, and a HW and SW architecture for such a modem is proposed. The HW architecture of an IEEE 802.16 compliant modem, which is depicted in Figure 6.4, looks very similar. It consists of analog radio portions; digital accelerators for modulation, demodulation, and time-critical MAC functions; and a programmable processor, on which the other MAC functions are executed in SW. For this processor, a 32-bit reduced instruction set computer (RISC) architecture, like ARM or LEON [6], in combination with a real-time operating system (RTOS), is the most popular choice.

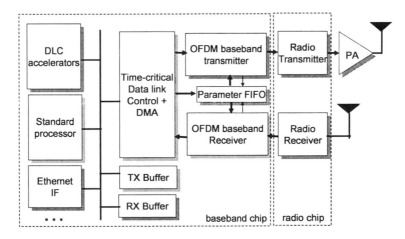

Figure 6.4. Top-level architecture of an IEEE 802.16 compliant modem.

One of the most critical aspects of the IEEE 802.16 modem architecture is the interaction between the MAC SW functions and the digital HW modules. The time-critical data link control (DLC) functions, including packing, fragmenting, cyclic redundancy check (CRC) inclusion and validation, processing of the downlink and uplink maps, and filtering of the received packets, are all performed in HW. Only the data communication and management packets that are intended for the client are transferred to the processor for further processing. The processor takes cares of the other DLC functions, the service-specific convergence sublayer and the higher-layer protocols. For computationally intensive tasks, like encryption, HW accelerators are often foreseen. The processor also serves as a bridge between the WiMax access network and the in-house Ethernet based network. Therefore, a HW Ethernet interface is often included.

Because of the different clocks of the HW and the processor, and the possible variations in processing latency in the processor, the communication between these two parts of the system must be buffered. Traditionally, a dual-port memory is used for this purpose. An alternative solution is the reservation of a part of the memory that can be accessed via direct memory access (DMA) for a transmission (TX) buffer for communication commands, and a reception (RX) buffer for received data packets. Remark that not only the data needs to be buffered, but also the associated parameters. To this end, the architecture of *Figure 6.4* includes a dedicated parameter first-in-first-out (FIFO) buffer.

The processor and the digital HW modules can be easily integrated on a single chip in complementary metal-oxide semiconductor (CMOS) technology. Only large memories are normally realized by means of external components. Due to the progress in circuit design over the last decades, also the radios can be made in CMOS technology. This enables the integration of the analog radio with the DSP portion, as demonstrated for WLAN in [7] and [8]. However, the IEEE 802.16 standard allows for much more variation in carrier frequencies, channel bandwidths, and duplexing schemes than WLAN. Consequently, the integration would require a very agile radio implementation, which is still an active research area. Therefore, all commercially available IEEE 802.16 compliant solutions currently still assume a two chip approach. Remark that the analog front-end in itself is also not necessarily monolithic, but consists of multiple chips. The power amplifier (PA), for instance, is often implemented in a different technology, and, thus, has to be realized with an external component.

Because the baseband chip should be able to interface with multiple analog front-ends, a flexible solution is needed. An example of such a flexible radio interface is depicted in *Figure 6.5*. The signal path consists of two analog-to-digital converters (ADCs) and two digital-to-analog converters (DACs), supporting direct up- and down-conversion. In addition to the signal DACs, digital and analog control outputs are foreseen, to set the frequency synthesizer parameters, the switches and the amplifications in the analog front-end. By means of a programmable digital waveform generator, a large variety of front-ends with varying control requirements can be supported. To enable feedback from the front-end in the form of, e.g., received signal strength indication (RSSI), digital and analog control inputs are also provided.

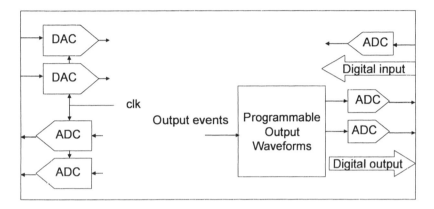

Figure 6.5. Flexible radio interface.

The base station has a very similar architecture for the PHY. However, as indicated before, the DLC and higher layer processing is much more demanding at the base station. Indeed, a base station should handle multiple users with high traffic volumes and low latency. Moreover, for a sectorized base station, the equipment is often requested to support multiple sectors. In most solutions, the increased requirements are addressed by means of a combination of multiple specialized network processors, for example the Intel IXP2350 processor [9], and a Gigabit Ethernet backhaul.

6.3 BASEBAND RECEIVER IMPLEMENTATION

In this section, we focus on the digital modulator and demodulator modules. First, Subsection 6.3.1 highlights the similarity between single-carrier and OFDM for multipath propagation environments. Next, Subsection 6.3.2 reviews some channel estimation approaches. Finally, Subsection 6.3.3 reconsiders some synchronization approaches.

6.3.1 Single-carrier versus OFDM

As motivated in Chapter 5, it can be expected that the OFDM scheme will become the dominant IEEE 802.16 PHY mode in the market. The efficient implementation of this OFDM mode can bootstrap on the extensive experience with the implementation of IEEE 802.11a and 11g WLAN modems, both in industry and academia. These designs have been extensively described in [7], [8], [10], and [11].

OFDM modems per definition use frequency-domain equalization. For single-carrier modems, time-domain equalization is traditionally used. However, to match the OFDM performance in multipath environments with highly frequency-selective channel responses, also the single-carrier mode should use frequency-domain equalization. Single-carrier modulation with frequency-domain equalization was initially introduced by Sari et al. in [12] and has been extensively studied since then. A comprehensive overview can be found in [13].

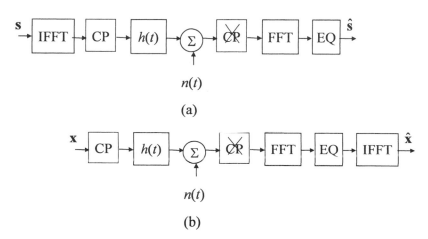

Figure 6.6. Frequency-domain processing of (b) single-carrier signals is closely related to (a) OFDM.

For frequency-domain processing of a single-carrier system, the received signal is first transformed towards the frequency-domain, and the equalized signal is next transformed back towards the time-domain. Hence, this scheme can be considered as an OFDM communication scheme, in which the IFFT of the transmitter is moved to the receiver. This relation is graphically illustrated in *Figure 6.6*. In the OFDM scheme, a signal block s in the transmitter is first mapped onto a number of parallel subcarriers, and converted to the time-domain by means of an IFFT operation. Next, a cyclic prefix (CP) is added and the signal is transmitted over the propagation channel with transfer function $h(t)$ and additive noise $n(t)$. At the receiver, the CP is discarded, the signal is converted to the frequency-domain by means of an FFT operation, and the equalization is performed on a per subcarrier basis, resulting in an estimate of **s** that is represented as $\hat{\mathbf{s}}$.

In the frequency-domain single-carrier scheme, the operation of the receiver is identical up to the point, where the equalized signals are converted back from the frequency- to the time-domain, to obtain the estimation of the original signal. The operation of the transmitter is limited to the addition of a CP to the time-domain symbol **x**.

Indeed, in the same way as for OFDM, the single-carrier waveform needs to have a cyclic extension to accommodate the delay spread of the propagation channel. To this end, the single-carrier signal must be divided in blocks. As illustrated in *Figure 6.7*, two ways exist to realize the cyclic property for these blocks:

1) One method is to define a CP, which consists of a copy of the end of the block that is prepended at the beginning of the block. The FFT at the receiver side is, hence, taken over the user data in the block.

2) The second way is based on the repetitive inclusion of an identical sequence of training symbols [14]. The FFT is consequently taken over the user data and one training sequence. Consequently, for the same FFT-size, the overhead of the cyclic extension is larger. On the other hand, the inclusion of specific training sequences allows them to be also used for additional purposes, such as channel estimation and/or symbol synchronization.

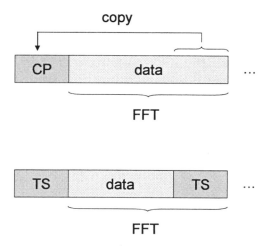

Figure 6.7. Frequency-domain processing can be realized with two different single-carrier waveforms, based on cyclic prefix (CP) or training sequences (TS).

The single-carrier mode of the IEEE 802.16 standard uses the unique training symbols in normal operation, and the cyclic prefix for some specific communication schemes, like, for instance, in combination with space-time coded modulation. Although the inclusion of these training symbols enables frequency-domain equalization of the received signal, this is not mandated by the standard. However, to reach an acceptable performance in NLOS conditions with severe multipath propagation, it is highly recommended. Therefore, we will restrict our attention in the remainder of this chapter towards frequency-domain equalization, both for OFDM and single-carrier.

The performance of OFDM and single-carrier with frequency-domain equalization has been studied by several authors [13][15][16]. From these studies, it can be concluded that single-carrier outperforms OFDM for high coding rates. This can easily be understood by the fact that OFDM requires a significant amount of FEC coding to exploit the inherent frequency diversity, which is not the case for single-carrier modulation. For lower coding rates, OFDM slightly outperforms single-carrier modulation. However, for single-carrier modulation, the frequency-domain equalizer can be complemented with a decision feedback equalizer (DFE). This DFE improves the performance, making that block-based single-carrier has a small performance advantage over OFDM for low coding rates.

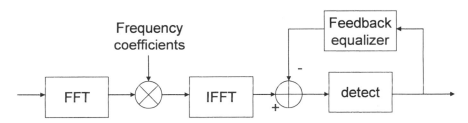

Figure 6.8. A block-based single-carrier system also allows to use a decision feedback equalizer (DFE).

The above performance comparison has been performed under idealized conditions. However, the inclusion of real-life constraints has a dramatic influence on the conclusions, as illustrated by [17], where PAs with limited linearity are included in the comparison. As can be expected, limited PA linearity, favors single-carrier modulation, especially with lower order modulation formats. On the other hand, the comparison does not take into account that OFDM could exploit adaptive per subcarrier modulation. This would improve its performance in heavy multipath environments, and make it very robust against inference.

As shown before, the basic scheme for a block-based signal-carrier implementation is very similar to an OFDM realization. Moreover, other receiver functions, like synchronization and channel estimation, are based on the block-based structure of the signal, and, hence, are very similar for both modulation schemes. It has also been proven that the sensitivity of block-based single-carrier and OFDM to carrier frequency offsets, phase noise, quadrature mismatch, and timing offsets is almost identical [17][18]. Therefore, we will not make a distinction between the two systems in the subsequent sections. If not explicitly mentioned, the proposed techniques will be compatible with any block-based modulation format.

6.3.2 Channel estimation

Frequency-domain equalization relies on the knowledge of the attenuation and the phase rotation that are caused by the channel on each (flat fading) subcarrier. Channel estimation is the process of obtaining these complex channel coefficients. Initial channel estimation that is performed at the start of each communication is called channel acquisition. When the communication uses long packets, or the communication is scheduled at regular intervals, it is useful to adapt the channel estimates continuously. With this so-called channel tracking, temporal variations as well as small frequency and time errors can be compensated. Most channel estimators are first derived for the non-adaptive case, and, subsequently, extended to the adaptive case.

Channel acquisition and tracking can be based on known training pilots in the time- or frequency-domain, or they can be performed in a blind manner, that is, without a-priori knowledge about transmitted symbols. Pilot-based estimation techniques are common in most communication systems, in which the sender emits some known signal to estimate the channel [19][20]. Two types of training symbols can be considered: pilot blocks and pilot subcarriers. A pilot block is a standard time-domain block of the modulation scheme with known data. Pilot subcarriers can only be used for OFDM transmission, and comprise typically a small fraction of the subcarriers, on which known data is transmitted. Usually these pilot subcarriers are evenly spaced over the complete frequency band. The pilots can either remain on fixed positions, or they could have varying positions from one OFDM symbol to the next. In the latter case, rotating pilots that sequentially step through all available pilot positions are most often used.

The simplest form of blind channel tracking consists of using the estimation of the transmitted signal as a virtual pilot carrier [21]. For single-carrier block transmission, the decision is taken in the time-domain and, hence, must be converted back to the frequency-domain by means of an FFT operation (see *Figure 6.9*).

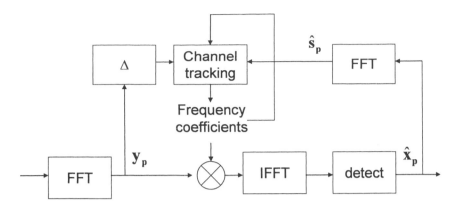

Figure 6.9. Channel tracking for single-carrier block transmission by means of virtual pilot
subcarriers.

Blind channel acquisition, on the other hand, only relies on the inherent features, or properties, of the transmitted signals. These features can be deterministic in nature, e.g., their so-called Finite Alphabet (FA) property or constant amplitude property, or they can be statistical in nature, like the cyclo-stationarity. Although blind estimation has been extensively studied in academic references [22][23][24][25], it is seldom used as the main technique in practical block-based systems, because of its slow convergence. On the contrary, most practical implementations use pilot-based techniques for acquisition, and a combination of pilot-based and blind estimation for tracking.

The simplest channel estimator for a block-based transmission consists simply in dividing the received signal in the frequency-domain, $\mathbf{y_p}$, by the frequency-domain symbols that have been actually sent $\mathbf{s_p}$. This estimator is usually known as the Least-Squares (LS) estimator, and can be written as:

$$\mathbf{h}_{LS} = \mathbf{y}_p / \mathbf{s}_p \tag{6.1}$$

where the division sign means element-wise division of $\mathbf{y_p}$ by $\mathbf{s_p}$. If a time-domain pilot symbol is available, all channel coefficients are estimated. In the other case, a decision-feedback mechanism can be used to estimate the channel on the data subcarriers.

The main advantage of the LS estimator is its simplicity, namely, one complex division per subcarrier. Its main disadvantage is the poor performance due to the use of an oversimplified channel model. Indeed, the LS estimator is based on the parallel Gaussian channel model. As a consequence, the frequency- and time-correlation of the channel are not taken into account in the LS estimator.

To improve the performance of the LS estimators, several methods were proposed that exploit the frequency-correlation of the channel during the estimation process. A first one in [26] is a linear minimum mean-squared error (LMMSE) estimator that minimizes the mean-squared error between the real and estimated channel coefficients. By doing so, the correlation between the channel coefficients on neighboring subcarriers is implicitly taken into account.

A second improved estimator is based on the observation that this frequency-correlation between channel coefficients is caused by the limited length of the channel impulse response. A Maximum Likelihood (ML) estimator that is based on this limited channel length property is proposed in [27]. Apart from its optimality in the ML sense (under the channel model hypothesis), this method, somewhat surprisingly, leads to a low complexity implementation, as illustrated in *Figure 6.10*. Remark that the truncation of the channel impulse response can be effectively implemented in the time-domain. Alternatively, an equivalent low pass filtering can be performed in the frequency-domain.

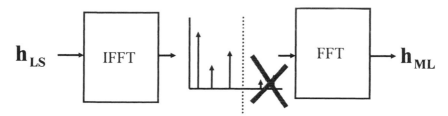

Figure 6.10. Limited length of channel impuls response can be exploited in the channel estimation process.

The combination of pilot blocks and pilot subcarriers theoretically allows a channel estimator to exploit both time- and frequency-correlation. Such a two-dimensional estimator structure is generally too complex for a practical implementation. To reduce the complexity, separating the use of time- and frequency-correlation has been proposed in [28], still with prohibitive complexity.

All the above channel estimators do not rely on a specific channel model. Parametric channel estimation, on the other hand, is based on a specific channel model. It determines the parameters of this model, and infers the quantities of interest (for example the frequency response) from this parametric model [29]. Parametric channel estimation usually offers better performance, since the number of quantities to be estimated is smaller. On the other hand, it potentially suffers from model mismatch problems.

6.3.3 Synchronization

An IEEE 802.16 modem assumes coherent detection, which means that the carrier frequency and symbol timing of the transmitter and receiver should be synchronized before actual user data can be exchanged. For a centrally scheduled MAC protocol, the initial frequency and symbol timing acquisition of a user terminal is the most challenging aspect of this synchronization. To this end, the downlink preambles at the start of each frame are used. The preamble structures for the WirelessMAN-SCa and WirelessMAN-OFDM PHYs show a large degree of similarity and are shown in *Figure 6.11*. The WirelessMAN-SCa downlink preamble starts with a ramp up (RU) section that contains a copy of the last part of a unique word (UW), followed by a repetition of unique words, which are random sequences of 16, 64, or 256 symbols. For WirelessMAN-OFDM, the downlink preamble consists of two OFDM symbols, each with a regular CP. The first one only contains non-zero data on every fourth subcarrier, resulting in four equal parts in the time domain. The second OFDM symbol only contains data on the even subcarriers, which leads to two identical halves. This repetition, which is present in both preamble structures, will be the essential property to estimate the carrier frequency and symbol timing. Because these preambles are broadcasted at the start of every downlink frame, multiple preambles could be used to improve the estimation. However, in the remainder of this section we will restrict our discussion to estimates based on a single preamble.

Once a first estimation of the carrier frequency and symbol timing is obtained, the subsequent downlink communication only requires slow tracking of these parameters. The downlink preambles could also be used to this end. An alternative is to exploit the cyclic block structure of the regular data communication symbols.

The user terminal will only start transmitting after acquisition of carrier frequency and symbol timing. Remark that for an FDD system, the used terminal will apply a fixed frequency offset to the acquired downlink frequency to determine the uplink frequency. As a consequence, no carrier frequency and symbol timing acquisition is needed at the basestation, but a slow tracking function will suffice. The cyclic block structure of the regular data communication symbols is normally used for this purpose.

In the remainder of this section, techniques for acquisition and tracking of the symbol timing misalignment and the carrier frequency offset will be presented. We categorize symbol timing algorithms in (a) methods based on the auto-correlation of the received signal, (b) methods based on the cross-correlation of the received signal with the transmitted pilot signal, and (c) methods based on the use of the cyclic prefix or known training sequence.

Where the first two methods are mostly used for symbol timing acquisition, the last one is mainly applicable for symbol timing tracking purposes. For carrier frequency offset estimation, the options are restricted to either methods based on auto-correlation, or methods based on the use of the CP or known training sequence.

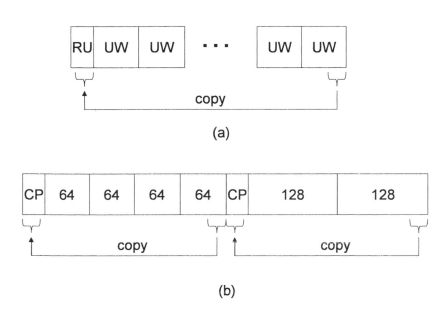

Figure 6.11. Burst preamble structure for (a) WirelessMAN-SCa and (b) WirelessMAN-OFDM PHYs.

6.3.3.1   Symbol timing synchronization

Before performing channel estimation, frequency-domain equalization, and demodulation, the correct timing of the block has to be acquired. Perfect synchronization is achieved if the data block selected for the FFT corresponds exactly to the transmitted data block. This is illustrated in *Figure 6.12*.

The effect of a mismatch in symbol timing is different for early and late synchronization. In the case of late synchronization, i.e., when the synchronization tick is later than the perfect synchronization point, a block of received data is selected that contains part of the CP of the next symbol, leading to inter-symbol interference (ISI).

In the case of early synchronization, the selected received data block contains part of the CP of the current block. As long as the sum of the timing offset and the maximum delay spread (expressed in number of samples) is smaller than the CP length, this results in perfect reception. If the offset is larger, ISI again appears.

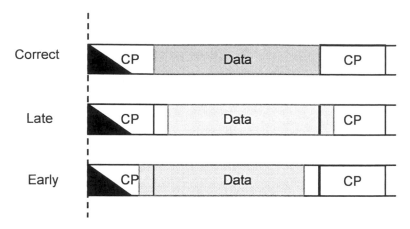

Figure 6.12. Correct, early, and late synchronization.

Timing synchronization techniques can be divided in cross-correlation methods, auto-correlation methods, and CP-based methods. The first two classes of methods have a higher accuracy, but also require a pilot symbol. This is not needed for the methods based on the CP. As a consequence, they are most often used for tracking of the symbol timing offset, while the acquisition is performed by a cross-correlation and/or auto-correlation method.

Cross-correlation methods [30] are used for symbol timing acquisition, and rely on the usage of a special pilot block $\mathbf{x_p}$ with N_p symbols. The receiver performs a correlation between the received time-domain pilot signal $\mathbf{r_p}$ and a local copy of the transmitted time-domain pilot block $\mathbf{x_p}$, according to the following equation:

$$\arg \max_{1 < \Delta n < N_p} \left(\sum_{n=1}^{N_p} r_p[n + \Delta n] \cdot x_p^*[n] \right) \qquad (6.2)$$

If the pilot sequence is well chosen, the output of the cross-correlation exhibits a dominant sharp peak in the case of an AWGN channel. However, as illustrated in *Figure 6.13*, multipath propagation will result in multiple peaks following the impulse response of the channel. With N_p^2 complex multiplications, cross-correlation methods incur a high computational load. Moreover, they require that the carrier frequency offset has been compensated before. This requirement resulted in a large number of proposals for combined symbol timing and frequency offset estimation [31].

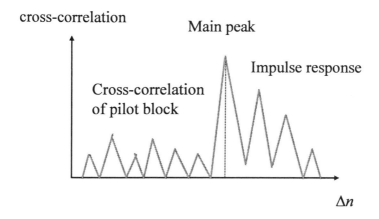

Figure 6.13. Cross-correlation in the presence of multipath propagation.

An alternative symbol timing acquisition method is based on the auto-correlation of the received pilot signal r_p. The most popular auto-correlation method is probably the so-called Schmidl and Cox method [32], which relies on a time-domain pilot block of length N_p with two identical halves in the time-domain. This pilot block is extended with a CP under multipath propagation conditions. The estimator takes the normalized auto-correlation between two parts of the received signal $\mathbf{r_p}$, separated by $N_p/2$ samples, and can be expressed by the following equation:

$$\arg_{1<\Delta n<N_p}\max \left(\frac{\sum\limits_{t=\Delta n+1}^{\Delta n+N_p/2} r_p^*[n]r_p\left[n+\frac{N_p}{2}\right]}{\sum\limits_{t=\Delta n+1}^{\Delta n+N_p/2} \left|r_p\left[n+\frac{N_p}{2}\right]\right|^2} \right) \qquad (6.3)$$

The main advantage of this metric is its robustness against multipath propagation and carrier frequency offset, due to the auto-correlation of two identically distorted signals. Furthermore, the normalization of the metric makes it robust against varying signal attenuations. The main disadvantage of the Schmidl and Cox method is the plateau-like timing metric, which gives rise to a relatively high uncertainty on the starting time of the symbol. This phenomenon is illustrated in *Figure 6.14.*

Auto-correlation

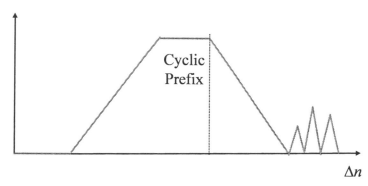

Figure 6.14. Auto-correlation shows a plateau, causing uncertainty about starting time.

For tracking purposes an auto-correlation method could also be applied on the CP of length N_{cp}, which is present in block-based transmission schemes [33]. Slow drift of the block timing is caused by the clock offset between the transmitter and receiver. In the case of an AWGN channel, the ML symbol timing estimator can be expressed as a function of the received data signal **r** and the *SNR* as follows:

$$\varepsilon_{ML} = \arg\max_{\varepsilon}\left(\left|\gamma(\varepsilon)\right| - \frac{SNR}{SNR+1}\,\Phi(\varepsilon)\right)$$

$$\text{where } \gamma(\varepsilon) = \sum_{n=\varepsilon}^{\varepsilon+N_{cp}} r[n]\, r^{*}[n+N_{cp}]\qquad\qquad (6.4)$$

$$\text{and }\quad \Phi(\varepsilon) = 0.5 \sum_{n=\varepsilon}^{\varepsilon+N_{cp}} \left|r[n]\right|^{2} + \left|r[n+N_{cp}]\right|^{2}$$

The main disadvantage of this estimator is its performance degradation in the presence of severe multipath propagation. Indeed, in the presence of multipath propagation, the range, on which the signal is cyclic, becomes smaller, and ultimately decreases to zero, when the channel delay spread becomes as long as the CP.

6.3.3.2   Carrier frequency synchronization

The acquisition of the carrier frequency is mostly achieved by means of an auto-correlation method. The first auto-correlation method was the so-called Moose method [34], in which a pilot symbol is transmitted that consists of a CP followed by two identical blocks of duration T_s. This method relies on the fact that the corresponding received blocks are related by the following expression, assuming a frequency offset Δf_c and a noiseless transmission:

$$\mathbf{r}_{l+1} = \mathbf{r}_l e^{j2\pi\Delta f_c T_s} \tag{6.5}$$

From this expression, one can easily show that the ML estimator of the carrier frequency offset is given by:

$$\hat{\Delta f_c} = \frac{1}{2\pi T_s} \tan^{-1}\left\{ \frac{\sum \mathrm{Im}(\mathbf{r}_{l+1}\mathbf{r}_l^*)}{\sum \mathrm{Re}(\mathbf{r}_{l+1}\mathbf{r}_l^*)} \right\} \tag{6.6}$$

Intuitively, this result is in accordance with the fact that the angle of $\mathbf{r}_{l+1}\mathbf{r}_l^*$ is equal to $2\pi \cdot \Delta f_c \cdot T_s$. For small frequency offsets, the tangent can be approximated by its argument, resulting in a simple estimator.

The main disadvantage of this method is its limitation to a carrier frequency offset that is smaller than $\pm 1/T_s$. Several authors have proposed solutions for this problem, based on pilot symbols with specific structures in the frequency-domain [31].

For carrier frequency offset tracking, methods based on the use of pilot subcarriers, or on the CP have been considered. In most OFDM schemes, a number of pilot subcarriers are foreseen, on which known data symbols are transmitted. When a residual carrier frequency offset is present, the individual subcarriers experience a constant phase rotation that is proportional to the carrier frequency offset. This phase error can be estimated from the received data on the pilot subcarriers. The subcarrier tracking mechanism is normally implicitly incorporated in the equalizer process. Alternatively, the phase error can also be estimated from the data subcarriers by means of a decision feedback scheme.

As an alternative to the use of pilot subcarriers, also the CP can be used to estimate the carrier frequency offset. This will result in an approach that is equivalent to ML estimation. As for the symbol timing estimator based on the CP, the main advantages of this estimator are its simplicity and the absence of pilots. Again, it is less robust against multipath propagation than auto-correlation methods. Similar to the approach for symbol timing synchronization, a known training sequence can be used, instead of the CP.

6.4 RADIO FRONT-ENDS

Signals generated by the digital modem at the transmit side of a BFWA system must be transported to the receive side of this system with minimum deterioration of the signal quality. Local authorities. like the FCC in the US, regulate this transmission by setting boundary conditions in terms of transmitted power, unintentionally radiated powers, and correct use of the available spectrum. The transmitter front-end and the PA transform the original baseband signal, so that it can be transported without violating these rules in the allocated frequency band. The receiving front-end downconverts, and filters the received signal with minimum impact on the SNR.

Traditionally, radio designers aimed at the conception of excellent radios with a high linearity, but also resulting in a high complexity, cost, and power consumption. More recently, the trend has changed to simpler albeit "dirtier" radios, whose analog impairments are compensated by carefully designed DSP algorithms. In the next sections, we will discuss some of these digital compensation techniques that are relevant to BFWA systems. Because of their different nature, we will separately treat the PA issues, which are discussed in Subsection 6.4.1, and the analog radio impairments, which are discussed in Subsection 6.4.2

6.4.1 Peak-to-average-power ratio reduction

In the desire to move to more bandwidth efficient modulation techniques, higher-order constellations are used, leading to larger variations in the amplitude of the baseband signal. This property of the signal is quantified by means of the peak-to-average-power ratio (PAPR), which is the ratio between the peak power in the signal and the average power in that same signal. For a signal $x[n]$ the PAPR can be expressed as:

$$PAPR = \left. \left(|x[n]|^2 \right)_{max} \middle/ E\left\{ |x[n]|^2 \right\} \right.$$

(6.7)

Instead of the PAPR, many authors consider the crest factor, which is the square root of the PAPR. Remark that in the above definitions, we used the discrete time-domain signal $x[n]$. However, the high crest factor not only has an impact on the number of bits for the digital implementation, but also has a major impact on the linearity of the PA. For the latter effect, the crest factor of the continuous time signal is needed. A typical procedure to obtain a good estimate of this continuous time crest factor consists of the upsampling and interpolation of the discrete time-domain signal, and the subsequent calculation of the crest factor of this upsampled signal.

Although single-carrier modulation with root raised cosine filtering will experience increasing PAPR with higher-order constellations, especially, OFDM modulation will suffer from this effect. However, for OFDM the large signal peaks will happen less frequently.

Signals with high PAPR are very sensitive to non-linear distortion, as the signal is amplified all along the transmit chain, and, particularly, in the PA. This non-linear distortion generates out-of-band harmonics, which have to be filtered out before transmission, as well as intermodulation products falling in the transmission bandwidth, which cause performance degradation.

A method that limits the detrimental influence of a large crest factor, rather than preventing the effect itself, is the clipping of the baseband signal in the digital domain. For OFDM, the clipping approach is well documented in literature [35], and it has been shown that it results in acceptable implementation losses.

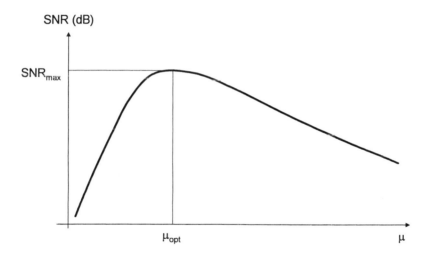

Figure 6.15. The optimum clipping level depends on the word length of the transmitted symbols.

The impact of digital clipping should not be considered separately from the effect of limited word lengths of the signal representation in the digital domain. This holds especially for the word length b of the symbol at the output of the transmitting digital modem. This word length has a major impact both on the implementation cost and the modem performance. As b decreases, the power consumption and the complexity of the DACs decreases at the expense of increased quantization noise. Quantizing and clipping both generate additive noise. Apart from their powers, these noise sources are uncorrelated. For a given word length of the samples of the baseband signal, reducing the normalized clipping level μ, increases the clipping noise, but at the same time reduces the quantization noise. In *Figure 6.15*, the SNR of a signal after clipping and quantizing is plotted versus μ for a given word length *b*. The clipping level μ is normalized with respect to the variance of the signal.

In the case of a complex baseband signal, the clipping can be performed on the in-phase (*I*) and quadrature (*Q*) signals separately. However, the signal that forms the input to the PA is the corresponding real passband signal. As a consequence, clipping on the magnitude of the complex signal (*I+j.Q*) is a better alternative. In *Figure 6.16*, clipping on the real and composite signals is compared. The shaded area represents the small difference between the two clipping strategies.

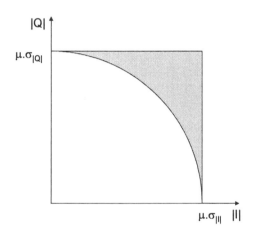

Figure 6.16. Clipping on the magnitude of I and Q separately, or on the magnitude of |I+jQ|.

This magnitude clipping operation can be implemented at low cost if the transmitter architecture includes a digital IF (with digital I/Q modulation). If not, it requires a considerable amount of DSP, hence increasing the system cost and power consumption. During system design, this must be traded off against the PA implementation cost.

The amount of clipping noise depends on the probability that the signal is clipped. A large number of techniques have been proposed to reduce this probability of clipping for OFDM signals, and, hence, reduce the effect of the crest factor. In [36], Schurgers et al. present a nice overview of these techniques. They made a distinction between two major approaches: block coding schemes and probabilistic methods. Block coding schemes aim at limiting the maximum amplitude of **x**, at the cost of a lower data rate. Ideally, these schemes completely eliminate the possibility of clipping. Probabilistic methods, on the other hand, are non-deterministic, as block coding schemes, but they reduce the clipping noise directly by lowering the probability of clipping. However, most of these techniques require a change to the standard transmission format, and, hence, cannot be applied in practice.

6.4.2 Digital compensation of analog impairments

Signals generated by the digital modem at the transmitter side of a wireless communication system are put on a specific carrier frequency to be transported to the receiver side. This is done by the transmitter front-end. Local authorities, like. the FCC in the US, regulate this transmission by setting boundary conditions in terms of unintentionally radiated powers, and correct use of the available spectrum. In addition, the standards define some desired properties of the transmitted signal. At the receiver side, the signal is down-converted by the receiver front-end, before it can be processed by the digital baseband modem.

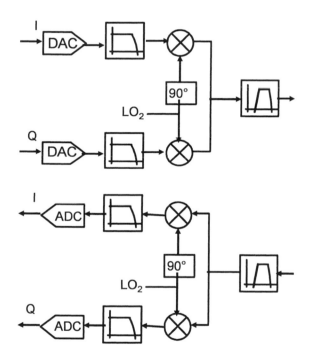

Figure 6.17. Direct conversion transceiver front-end architecture.

Because of the large variety in BFWA standards, also many possible architectures exist for the front-end [37]. A distinction has to be made between time-domain duplexing (TDD), frequency-domain duplexing (FDD), or half frequency-domain duplexing (HFDD) solutions. For each of these cases, intermediate frequency (IF) or direct conversion architectures are possible. Because the direct conversion architecture, as shown in *Figure*

6.17, is the most cost effective one but also the most challenging to the digital modem, we will focus the rest of this section to this architecture.

In a direct conversion architecture, the digitally modulated signal is first transformed into an analog signal, and then up-converted in a single step to the assigned radio frequency. Furthermore, the up-conversion is combined with I/Q modulation. Each operation is followed by a filtering operation to avoid unintentional transmission in adjacent frequency bands. At the receive side, the signal undergoes similar operations in reverse order, from the receiver input (antenna or cable connector) to the ADCs. The first operation always consists of the separation of the wanted signal from blockers and signals in adjacent frequency bands by means of filtering.

Both at the transmit and receive side, the modulated signal will be impaired by several non-ideal processing steps in the analog front-end:

- First, gain and phase mismatches between the I and Q paths in the I/Q (de)modulator, lead to cross talk between the I and Q signals and to distortion of the constellation;
- Second, phase noise on the local oscillator port of the mixers, causes inter-(sub)carrier-interference (ICI) in OFDM. Block-based single-carrier systems are less affected by phase noise, although special care must be taken during the estimation of the frequency-domain channel coefficients;
- Finally, the channel impulse response is extended by means of the impulse response of the transmit and receive filters. This effect causes ISI, in case the resulting composite impulse response exceeds the length of the CP or known training sequence.

These effects not only affect the data symbols but also the preamble symbols that are used for estimation of the propagation channel parameters. As a consequence, they can have a detrimental impact on the modem performance. For each of these effects, digital compensation techniques have been proposed. In the remainder of this subsection, we will give a brief overview of these methods.

6.4.2.1   I/Q mismatch

The mismatch between the I and the Q branch of a front-end can take two forms. On the one hand, the gain of both branches can be different. On the other hand, the phase difference between the I and Q branch can deviate from 90°. When distributing these gain and phase errors equally over the two branches, and defining the distortion-free complex received baseband signal as $\mathbf{r}$ and the distorted signal as $\mathbf{r_{iq}}$, we can express the effect of I/Q mismatch as follows:

$$\mathbf{r_{iq}} = \alpha.\mathbf{r} + \beta.\mathbf{r}^* \qquad (6.8)$$

For small errors, α is close to 1 and β is approaching 0. As a consequence, it can be concluded that the effect of I/Q imbalance on the time-domain signal consists in the scaling of the wanted signal by a complex scaling factor, followed by the superposition of a heavily attenuated version of its complex conjugate. In the case of a block-based transmission, the effect of I/Q mismatch in the frequency-domain can be calculated by taking the FFT of a block of $\mathbf{r_{iq}}$, resulting in:

$$\mathbf{y_{iq}} = \alpha.\mathbf{y} + \beta.\mathbf{y}_m^* \qquad (6.9)$$

where $\mathbf{y_m}$ represents the mirrored version of $\mathbf{y}$, obtained by reversing the indices of the vector of frequency-domain coefficients. From this equation, we learn that I/Q mismatch has two effects on the received frequency-domain vector: scaling of the desired signal with a complex factor α, and addition of a scaled and mirrored version of this desired signal.

Tubbax et al. have compared the performance and sensitivity of OFDM and block-based single-carrier modulation to I/Q mismatch [38]. Although their results have been derived for a WLAN set-up with 64 carriers, it can be expected that the results are also valid for a BFWA system with 256 carriers. Their simulations in an AWGN propagation channel show that both schemes are equally sensitive.

When the I/Q mismatch is known, i.e., the scaling factors α and β, the effect of the distortion can easily be compensated in the digital domain. Indeed, starting from (6.8), we can easily derive that:

$$\hat{\mathbf{r}} = \frac{\alpha^*.\mathbf{r_{iq}} - \beta.\mathbf{r_{iq}}^*}{|\alpha|^2 - |\beta|^2} \qquad (6.10)$$

with $\hat{\mathbf{r}}$ the estimation of $\mathbf{r}$. Hence, compensating the I/Q mismatch boils down to an accurate estimation of α and β.

Recently several estimation approaches have been proposed, especially for the case of OFDM transmission. Fettweis et al., for instance, have worked out a blind estimation technique in [40]. In [39], several schemes (LMS, adaptive equalizer, pre-FFT) have been compared. Tubbax et al. have proposed a compensation algorithm for I/Q mismatch that is based on the property that the propagation channel features a non-zero coherence bandwidth, and, hence, should demonstrate a smooth behavior [41]. Because of the addition of a scaled and mirrored version of the frequency coefficients, a signal with sharp transition is superimposed. This effect is illustrated in *Figure 6.18*. Hence, these sharp transitions can be used for estimating α and β. During the evaluation of the algorithms, it was proven that they work fine as well in the presence of small frequency offsets or phase noise.

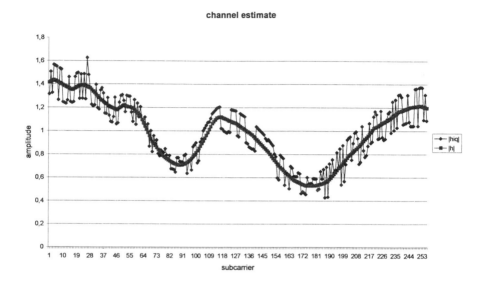

Figure 6.18. The effect of I/Q imbalance on channel estimation.

6.4.2.2 Phase noise

Phase noise is of primary concern to OFDM systems, where it has two effects on an OFDM symbol, namely, the Own Noise Contribution (ONC) and the Foreign Noise Contribution (FNC). The ONC comes from the low-frequency part of the phase noise. It results in an identical rotation of all subcarriers, and is therefore also called the Common Phase Error (CPE). The FNC comes from the high-frequency phase noise contributions. On each subcarrier, it gives rise to ICI from all other subcarriers.

Because the ONC causes an identical phase rotation for all subcarriers in an OFDM symbol, it can be estimated and corrected. A two-step approach based on decision feedback can be pursued. First, the phase rotation is estimated as the average angle between the received symbols and the hard decisions of these symbols. Next, the estimated phase rotation is applied to the OFDM symbol and the symbol decisions are redone. This approach eliminates the phase rotation caused by the ONC. The residual degradation that is left is caused by the FNC, which cannot be compensated for.

The efficiency of phase noise compensation is a function of the ratio between the ONC and the FNC. For BFWA deployments with large channel bandwidths, the inter-carrier spacing is large compared to the phase noise bandwidth, and the ONC becomes the dominant effect. For smaller channel bandwidths, the inter-carrier spacing is reduced and the FNC, which cannot be compensated for, becomes the main phase noise contribution.

6.4.2.3   Filtering

Traditionally, the specifications of a radio filter are expressed in the frequency-domain by the in-band ripple and the stop-band attenuation. However, for block-based transmission schemes also the time-domain characteristics have a large impact on the performance.

In specific, the extension of the impulse response of the channel due to the insertion of these filters is a crucial factor. Indeed, in block-based schemes, the CP is used to prevent ISI between subsequent symbols. To guarantee this property, the CP should have a minimal length, equal to the maximal length of the channel impulse response. The introduction of the radio filters might cause the length of the composite impulse response to exceed the CP length, resulting in ISI.

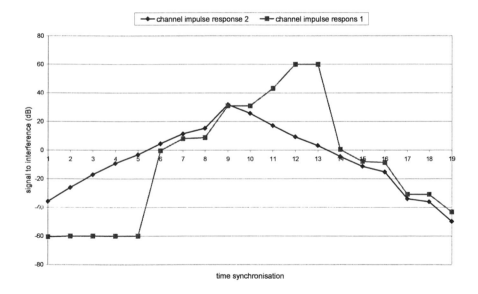

Figure 6.19. SIR versus synchronization location for two different multipath channels.

As pointed out in [42], the amount of ISI is also function of the symbol synchronization timing of the receiver. *Figure 6.19* shows the signal-to-interference ratio (SIR) as a function of the synchronization point for two different propagation channels. A 7-tap finite impulse response (FIR) filter and a CP of 16 samples are used. For the first channel response, a plateau-like behavior can be observed. For the second channel impulse response, there is a sharp peak around 30 dB SIR at the optimal synchronization time. A good filter design aims at a high SIR over at least a few samples for a large variety of multipath channels.

6.5 SUMMARY

In this chapter, we have provided a brief review of the implementation challenges in designing cost-efficient BFWA transceivers. The overall architecture was our first point of attention. The top-level architectural choices are normally taken at a very early stage in the design process, and may have a tremendous impact on the quality of the final result. Flexible interfaces between hardware and software and between digital and analog processing were pointed out as crucial features.

Next, the baseband receiver implementation has been discussed. It was demonstrated there is a large similarity between single-carrier modulation with frequency-domain processing and OFDM. Furthermore, several techniques for channel estimation, symbol timing synchronization, and carrier frequency synchronization, for these block-based communication schemes were briefly reviewed.

Finally, we discussed the radio front-ends. Our approach consisted in the adoption of a low-cost direct conversion radio and the compensation of the resulting analog impairments in the digital domain. In this perspective, we discussed techniques for PAPR reduction, I/Q imbalance compensation, phase noise compensation, and radio filter design.

6.6 REFERENCES

[1] N. Briscoe, "Understanding the OSI 7-Layer Model", PC Network Advisor, Issue 120, pp. 13-14, July 2000.

[2] M. Hawa, "Stochastic Evaluation of Fair Scheduling with Applications to Quality-of-Service in Broadband Wireless Access Networks" PhD dissertation, University of Kansas, August 2003.

[3] M. Engels (editor), "Wireless OFDM Systems: How to Make Them Work", Kluwer Academic Publishers, 2002.

[4] CableLabs, DOCSIS 2.0 Interface Specifications, http://www.cablemodem.com/.

[5] Y.-D. Lin, W.-M. Yin, C.-Y. Huang, "An Investigation into HFC MAC Protocols: Mechanisms, Implementation, and Research Issues", IEEE Communications Surveys, http://www.comsoc.org/pubs/surveys, Third Quarter 2000, pp. 1-13.

[6] LEON 3 processor, http://www.gaisler.com/

[7] S. Mehta, D. Weber, M. Terrovitis, K. Onodera, M. Mack, B. Kaczynski, H. Samavati, S. Jen, W. Si, M. Lee, K. Singh, S. Mendis, P. Husted, N. Zhang, B. McFarland, D. Su, T. Meng, B.Wooley, "An 802.11g WLAN SoC", IEEE ISSCC Digest of Technical Papers, Vol. 1,pp. 94-586, February 2005.

[8] H. Darabi, S. Khorram, Z. Zhou, T. Li, B. Marholev, J. Chiu, J. Castaneda, E. Chien, S. Anand, S. Wu, M. Pan, H. Kim, P. Littieri, B. Ibrahim, J. Rael, L. Tran, E. Geronaga, J. Trachewsky, A. Rofougaran, "A Fully Integrated SoC for 802.11b in 0.18 μm CMOS", IEEE ISSCC Digest of Technical Papers, Vol. 1,pp. 96-586, February 2005.

[9] M. Venkatachalam, "Integrated Data and Control Plane Processing Using Intel® IXP23XX Network Processors", Technology@Intel Magazine, http://developer.intel.com/technology/magazine/index.htm, February 2005.

[10] W. Eberle, et al., "A Digital 72Mb/s 64-QAM OFDM Transceiver for 5GHz Wireless LAN in 0.18um CMOS", IEEE ISSCC Digest of Technical Papers, pp. 336-337, February 2001.

[11] P. Ryan, et al., "A Single Chip PHY COFDM Modem for IEEE 802.11a with Integrated ADCs and DACs", IEEE ISSCC Digest of Technical Papers, pp. 338-339, February 2001.

[12] H. Sari, G. Karam, I. Jeanclaude, "Transmission Techniques for Digital Terrestrial TV Broadcasting", IEEE Communications Magazine, Vol. 33, No. 2, pp. 100-109, February 1995.

[13] D. Falconer, S. L. Ariyavisitakul, A. Benyamin-Seeyar, B. Eidson, "Frequency Domain Equalization for Single-Carrier Broadband Wireless Systems", IEEE Communications Magazine, Vol. 40, No. 4, pp. 58-66, April 2002.

[14] L. Deneire, B. Gyselinckx, M. Engels, "Training Sequence versus Cyclic Prefix - A New Look on Single-Carrier Communication", IEEE Communication Letters, Vol. 5, No. 7, pp. 292-294, July 2001.

[15] V. Aue, G. P. Fettweis, R. Valenzuela, "A Comparison of the Performance of Linearly Equalized Single-Carrier and Coded OFDM over Frequency Selective Fading Channels Using the Random Coding Technique, IEEE Proc. of ICC, Vol. 2, pp. 753-757, June 1998.

[16] N. Benvenuto, S. Tomasin, "On the Comparison Between OFDM and Single-Carrier Modulation With a DFE Using a Frequency-Domain", IEEE Transactions on Communications, Vol. 50, No. 6, pp. 947-955, Jun 2002.

[17] J. Tubbax, B. Côme, L. Van der Perre, L. Deneire, M. Engels, "OFDM versus Single-Carrier with Cyclic Prefix: a System-Based Comparison for Binary Modulation", IEEE Proc. of WPMC, pp. 537--540, September 2001.

[18] Z. Wang, X. Ma, and G. B. Giannakis, "OFDM or Single-Carrier Block Transmissions?", IEEE Transactions on Communications, Vol. 52, No. 3, pp. 380-394, March 2004.

[19] Y. Li, "Pilot-Symbol-Aided Channel Estimation for OFDM in Wireless Systems", IEEE Proc. of VTC, Vol. 2, pp. 1131-1135, May 1999.

[20] M. Morelli, U. Mengali, "A Comparison of Pilot-Aided Channel Estimation Methods for OFDM Systems", IEEE Transactions on Signal Processing, Vol. 49, No. 12, pp. 3065-3073, December 2001.

[21] S. Coleri, M. Ergen, A. Puri,.A. Bahai, "Channel Estimation Techniques Based on Pilot Arrangement in OFDM Systems", IEEE Transactions on Broadcasting, Vol. 48, No. 3, pp. 223- 229, September 2002.

[22] K. Abed-Meraim, W. Qiu, and Y. Hua, "Blind System Identification", Proc. of the IEEE, Vol. 85, No. 8, pp. 1310–1322, Aug. 1997.

[23] A. Scaglione, G. B. Giannakis, S. Barbarossa, "Redundant Filterbank Precoders and Equalizers Part II: Blind Channel Estimation, Synchronization, and Direct Estimation", IEEE Transactions on Signal Processing, Vol. 47, No. 7, pp. 2007–2022, July 1999.

[24] B. Muquet, M. de Courville, P. Duhamel, V. Buenac, "A Subspace Based Blind and Semi-Blind Channel Identification Method for OFDM Systems", IEEE Proc. of SPAWC, pp. 170–173, May 1999.

[25] R. W. Heath, G. B. Giannakis, "Exploiting Input Cyclostationarity for Blind Channel Identification in OFDM Systems", IEEE Transactions on Signal Processing, Vol. 47, No, 3, pp. 848–856, March 1999.

[26] O. Edfors, M. Sandell, J.J. van de Beek, S. K. Wilson, P. O. Borjesson, "OFDM Channel Estimation by Singular Value Decomposition", IEEE Transactions on Communications, Vol. 46, No. 7, pp. 931-939, July 1998.

[27] L. Deneire, P. Vandenameele, L. Van der Perre, B. Gyselinckx, M. Engels, "A Low Complexity ML Channel Estimator for OFDM", IEEE Proc. of ICC, Vol. 5, pp. 1461-1465, June 2001.

[28] P. Hoher, "TCM on Frequency-Selective Land-Mobile Fading Channels", in Proc. of Tirrenia Int. Workshop on Digital Communications, pp. 317–328, September 1991.

[29] B. Yang, K. B. Letaief, R. S. Cheng, Z. Cao, "Channel Estimation for OFDM Transmission in Multipath Fading Channels Based on Parametric Channel Modeling", IEEE Transactions on Communications, Vol. 49, No. 3, pp. 467-479, March 2001.

[30] W. D. Warner, C. Leung, "OFDM/FM Frame Synchronization for Mobile Radio Data Communication", IEEE Transactions on Vehicular Technology, Vol. 42, No. 3, pp. 302-313, August 1993.

[31] T. Keller, L. Piazzo, P. Mandarini, L. Hanzo, "Orthogonal Frequency Division Multiplex Synchronization Techniques for Frequency-Selective Fading Channels", IEEE Journal on Selected Areas in Communications, Vol. 19, No. 6, pp. 999-1008, June 2001.

[32] T. Schmidl, D. Cox, "Robust Frequency and Timing Synchronization for OFDM", IEEE Transactions on Communications, Vol. 45, No. 12, pp. 1613-1621, December 1997.

[33] J.-J. van de Beek, M. Sandell, P. O. Börjesson, "ML estimation of time and frequency offset in OFDM systems", IEEE Transactions on Signal Processing, Vol. 45, No. 7, pp. 1800-1805, July 1997.

[34] P. H. Moose, "A Technique for Orthogonal Frequency Division Multiplexing Frequency Offset Correction", IEEE Transactions on Communications, Vol. 42, No. 10, pp.2908-2914, October 1994.

[35] D.J.G. Mestdagh, P.M.P. Spruyt, et. al. , "Effect of Amplitude Clipping in DMT-ADSL Transceivers", Electronic Letters, Vol. 29, No 15, pp. 1354-1355, July 1993.

[36] C. Schurgers, M. B. Srivastava, "A Systematic Approach to Peak-to-Average-Power Ratio in OFDM", SPIE's 47th Annual Meeting, pp. 454-464, August 2001.

[37] B. Bisla, R. Eline, L. M. Franca-Neto, "RF System and Circuit Challenges for WiMAX", Intel Technology Journal, Vol. 8, No. 3, pp. 189- 200, August 2004.

[38] J. Tubbax, B. Côme, L. Van der Perre, L. Deneire, S. Donnay, M. Engels, "OFDM versus Single Carrier with Cyclic Prefix: a System-Based Comparison", IEEE Proc. of VTC-Fall, Vol. 2, pp. 1115-19, October 2001.

[39] A. Tarighat, R. Bagheri, A. H. Sayed, "Compensation Schemes and Performance Analysis of I/Q Imbalances in OFDM Receivers", IEEE Transactions on Signal Processing, Vol. 53, No. 8, Part 2, pp. 3257 – 3268, August 2005.

[40] M. Windisch, G. Fettweis, "Standard-Independent I/Q Imbalance Compensation in OFDM Direct-Conversion Receivers", 9th International OFDM Workshop, pp. 57-61, September 2004.

[41] J. Tubbax, B. Côme, L. Van der Perre, L. Deneire, S. Donnay, M. Engels, "I/Q Imbalance Compensation for OFDM", IEEE Proceedings of ICC, Vol. 5, pp. 3403-3407, May 2003.
[42] B. Debaillie, B. Côme, W. Eberle, S. Donnay, M. Engels, I. Bolsens, "Impact of Front-End Filters on Bit Error Rate Performances in WLAN-OFDM Transceivers", IEEE Proc. of RAWCON, pp. 193-196, August 2001.

Chapter 7

Adding a Dimension with Multiple Antennas
Overview of Smart Antenna Systems

Frederik Petré
With contribution of Nadia Khaled

7.1 INTRODUCTION

In Chapter 5, it was shown that the BFWA physical layer enables reliable data communication over the inherently hostile BFWA propagation channel by successfully coping with its main impairments, such as path loss, delay dispersion, time variance, and co-channel interference. However, in order to meet the data rate and Quality-of-Service (QoS) requirements of future broadband wireless services, the spectral efficiency and the link reliability of the BFWA system as a whole, and its PHY in particular, should be considerably improved, which can not be realized by traditional single-antenna communication techniques.

It is now widely believed, both in research and industrial communities, that smart multi-antenna communication techniques are key to meet the above challenges in terms of data rates and QoS. On the one hand, smart antenna systems at either the transmit or receive side, improve the link reliability in terms of receiver Signal-to-Interference-and-Noise-Ratio (SINR), by coherently combining multiple independent observations of the transmitted signal, while rejecting unwanted interfering signals. This is achieved through the so-called array, diversity, and interference reduction gains offered by smart antennas. On the other hand, Multi-Input Multi-Output (MIMO) antenna systems, which deploy multiple antennas at both ends of the wireless link, truly open an additional spatial dimension, besides the traditional time and frequency dimensions, to significantly increase the spectral efficiency (besides improving the link reliability), relative to single-antenna systems. Specifically, MIMO antenna systems create a number of parallel spatial pipes, through which independent information streams can be simultaneously transferred at the same frequency. This is referred to as the spatial multiplexing gain.

To obtain a rigorous classification of smart antenna systems, many different approaches can be pursued. In this chapter, we have opted for a taxonomy based on the location of the smart antenna processing, either only at the base station, or only at the subscriber station, or at both sides of the wireless link.

MIMO-OFDM transmission turns out to be of particular interest, since it synergistically combines the advantages of OFDM and MIMO communications. Apart from the sheer accumulation of the respective advantages of both techniques, the combination of OFDM, which elegantly mitigates delay dispersion and frequency-selective fading, with MIMO communication, which increases spectral efficiency and improves link reliability, results in a high implementation efficiency, by exploiting the inherent subcarrier parallelism. The potential of MIMO-OFDM in particular, and smart antenna systems in general, has also been recognized by the IEEE 802.16 Broadband Wireless Access Working Group, which explores different routes for the inclusion of such advanced communication techniques in the standard.

This chapter is organized as follows. Section 7.2 points out the potential gains of smart antenna systems, while Section 7.3 proposes a taxonomy of such systems, based on their system configuration. Section 7.4 explores the potential of MIMO-OFDM transmission, and reviews several MIMO-OFDM processing techniques, while Section 7.5 discusses the application of smart antenna techniques in the IEEE 802.16 standard. Finally, Section 7.6 draws some conclusions.

7.2 POTENTIAL GAINS OF SMART ANTENNAS

Through additionally exploiting the space dimension, smart antenna systems are able to capitalize on the rich multipath nature of the BFWA networks to acquire multiple independent observations of the transmitted signals [1][2]. These multiple observations are then coherently combined to improve the ability of the receiver to recover the transmitted information and/or to increase the number of simultaneous data streams. The former is achieved through exploiting the diversity benefit, array gain and/or interference rejection capability offered by smart antennas, whereas the latter corresponds to their spatial multiplexing gain. In the following, we briefly review each of these four potential benefits of smart antenna systems.

7.2.1 Diversity gain

To combat fading in wireless channels, smart antenna systems capitalize on the principle that the signals received from two or more uncorrelated antennas, e.g. in space, polarization or antenna pattern, will experience independent fading. Accordingly, if one antenna is experiencing a fading signal, it is likely that the other antenna will not, so at least one good signal can be received. Furthermore, the larger the number of available uncorrelated antennas, the higher is the so-called diversity order. A higher diversity order lowers the probability of experiencing a deep fade and improves the resulting BER performance, as shown in *Figure 7.1*.

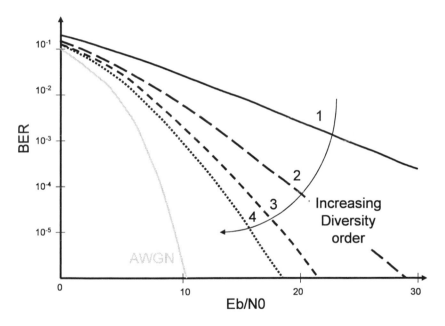

Figure 7.1. Diversity order improves BER performance.

7.2.2 Array gain

To extend the coverage of BFWA cells, smart antenna systems make use of the fact that multiple antenna receivers acquire multiple observations of the transmitted signal. If these observations are coherently combined, their corresponding received powers are constructively summed, leading to a gain in the received E_b/N_0. This gain is widely known as the receive array gain. Transmit array gain can be similarly defined for multiple-antenna transmitters.

An *M*-antenna transmitter or receiver provides a maximum array gain of *M*. The actually achieved array gain, however, critically depends on the availability of channel state information and the combining strategy.

7.2.3 Interference rejection gain

When strong interference is also present, diversity processing alone cannot improve the signal. To cope with interference, smart antennas can be utilized to shape the antenna radiation pattern in such a way as to enhance the desired signals and null the effect of the interfering signals. This is referred to as the interference rejection gain of smart antenna systems.

7.2.4 Spatial multiplexing gain

To extract the three previous smart antenna gains, it is sufficient to have multiple antennas at one side of the communication link. In this case, however, it is only possible to transmit a single data stream to guarantee data recovery. Alternatively, when multiple antennas are available at both the transmitter and the receiver, multiple independent data streams can be simultaneously transmitted. Provided the number of receive antennas is larger than the number of transmitted data streams, the latter can be perfectly recovered at the receiver. In fact, the matrix MIMO channel is equivalently decomposed into a number of spatial SISO sub-channels, over which multiple data streams are conveyed. Therefore, this transmission scheme is called spatial multiplexing, and the number of simultaneously transmitted streams is called spatial multiplexing gain.

It is important to note that an $M_T \times M_R$ MIMO system possesses $M_T M_R$ degrees of freedom, which can be exploited to extract one or several of the above introduced smart antenna gains. Nevertheless, constraining the system performance in terms of one of these gains automatically sets constraints on the achievable amount of the other gains.

7.3 TAXONOMY OF SMART ANTENNA SYSTEMS

There exist many ways to classify smart antenna systems. In this section, we propose a taxonomy based on their system configuration [3]. Thus, Subsection 7.3.1 considers smart antenna systems with multiple antennas at the base station only. Subsection 7.3.2 alternatively deals with systems with multiple antennas at the subscriber station only. Finally, MIMO systems, which use multiple antennas at both the base and subscriber station, are looked at in Subsection 7.3.3.

7.3.1 Base station antenna systems

In base station antenna systems, antenna processing is carried out only at the base station in either or both the up- and down-links. Because it concentrates processing and hardware complexity at the base station, this system configuration has enjoyed tremendous interest and popularity, and has actually been deployed in several existing cellular wireless systems.

7.3.1.1 Single-user case

To acquire the signal transmitted by the single-antenna subscriber station, the M_R-antenna base station can use three common diversity techniques. These are selection diversity, equal-gain combining and maximum-ratio combining. In selection diversity, only the antenna with the best signal is selected for reception. In equal-gain combining, the M_R received signals are co-phased and summed, for improved reception. Finally, maximum-ratio combining optimally combines the M_R antenna signals such that the resulting post-processing SNR is maximized. It is the optimum method in single-user AWGN channels, and achieves a full M_R-fold spatial diversity and an M_R-fold array gain. Obviously, both optimum selection and combining require knowledge of the channel, which the received signals have undergone.

Provided the channel is known at the base station prior to transmission, the three aforementioned uplink techniques can be similarly applied to the downlink with a normalization to meet the transmit power constraint. In transmit maximum-ratio combining, for instance, the M_T-antenna base station would optimally weigh the single transmit data stream across its antennas, such that channel filtering leads to maximum-SNR coherent reception at the single-antenna mobile terminal. However, channel state information may not be readily available at the transmitter (as extensively discussed in Subsection 7.4.3). In that case, space-time coding has been introduced, which through ingenious coding across the multiple transmit antennas and appropriate combining at the single antenna subscriber station, is able to exploit the available transmit spatial diversity [4][5][6]. A more detailed treatment of the space-time coding techniques can be found in Subsection 7.4.2.

7.3.1.2 Multi-user case

In multi-user scenarios, where multiple users can simultaneously access the wireless medium, the above-introduced single-user diversity techniques are not sufficient for successful reception. Dedicated diversity techniques are needed to either reject co-channel interferers of a particular user, or jointly detect co-channel users. In the following, we visit these two alternative scenarios for both up- and down-links.

Co-channel interference is present in all cellular systems. It is due to frequency re-use in TDMA and FDMA systems, while it is constitutive in CDMA systems. Smart antenna systems can help reject such co-channel interference. In fact, an M_R-antenna base station can capitalize on the spatial resolution provided by its multi-element antenna array, to discriminate the different users based on their distinct spatial signatures. Accordingly, it can shape its array radiation pattern in such a way as to enhance the desired user's signal and null the effect of the co-channel interferers [7][8], so-called optimum combining.

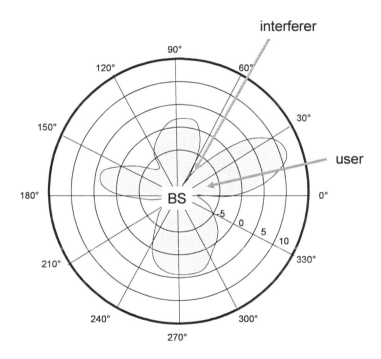

Figure 7.2. Smart antenna systems shape their array pattern to enhance the user's signal and reduce the interferer.

To enhance the network capacity, the base station may alternatively utilize its spatial discrimination capability to separate and detect the data streams corresponding to several co-channel mobile terminals, so-called Space Division Multiple Access (SDMA).

If the downlink channel state information is perfectly known at the base station, then transmit processing can be defined analogously to the uplink combining described for the single-user case. On the one hand, the smart antenna base station can deploy a transmit processing that equivalently steers a radiation beam in such a way as to maximize the signal at the desired user's antenna and minimize/null the induced interference at the co-channel mobile users. On the other hand, the smart antenna base station can alternatively deploy transmit SDMA processing to equivalently orthogonalize the data transmissions to the co-channel users.

7.3.2 Subscriber station antenna systems

Subscriber station antenna systems refer to systems with a single-antenna base station and a multiple-antenna subscriber station. Consequently, in these systems, antenna processing is performed only at the subscriber station in either or both the up- and downlinks. In principle, the methods, introduced in Subsection 7.3.1 for base station antenna systems, can be directly applied to subscriber station antenna systems. In the context of BFWA networks, however, subscriber station antenna systems are subject to two additional design constraints. First, the subscriber station must remain compact, low cost and with long-lasting battery life in case of a portable or mobile version. Second, in BFWA networks, each subscriber station is not aware of the channel state information related to its co-channel interferers. The first constraint imposes low-complexity antenna processing solutions, while the second excludes multi-user detection and transmission at the smart antenna subscriber station.

In the uplink, low-complexity space-time block coding [5][6] can be deployed to extract the spatial diversity offered by the M_T subscriber station antennas. However, provided the transmit channel knowledge can be obtained, using one of the methods proposed in Subsection 7.4.3, then the uplink transmission can further benefit from the available array gain. To do so, the subscriber station can deploy transmit maximum-ratio combining, or suboptimal transmit equal-gain combining and transmit antenna selection [9][10].

In the downlink, the subscriber station acquires the downlink channel state information. Based on that, it can either carry out selection diversity, equal-gain combining or maximal ratio combining to receive the single transmitted data stream. In the presence of co-channel users, it can further deploy optimum combining to suppress interference [7][8].

7.3.3 MIMO antenna systems

MIMO antenna systems possess smart antennas at both the base station and the subscriber station. Therefore, both stations are able to carry out antenna processing. This obviously translates into a larger flexibility in distributing the processing burden between the base station and the subscriber station. This larger flexibility has lead to a richness of processing techniques, which strike different diversity gain - array gain - multiplexing gain trade-offs, depending on their level of intelligence, complexity and a-priori channel knowledge.

A detailed overview of these techniques, and the scenarios where they are most suitable, is to be found in Subsection 7.4.2. A special emphasis is put on single-user scenarios, which are most appropriate for current BFWA networks. We do, however, provide pointers to their multi-user extensions.

7.4 MIMO-OFDM TRANSMISSION

This section first introduces OFDM-based MIMO transmission. It then provides an overview of the prominent state-of-the-art MIMO processing techniques, highlighting their respective transmit and receive channel state information requirements. Finally, this section more particularly examines the transmit CSI acquisition issue, which is of particular relevance for MIMO communication systems.

7.4.1 MIMO-OFDM system model

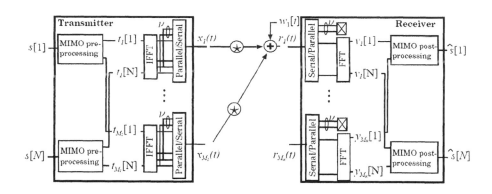

Figure 7.3. A generic MIMO-OFDM communication system.

Capitalizing on the SISO-OFDM system model previously introduced in Chapter 5, we herein describe its extension for OFDM-based MIMO systems. For an $M_T \times M_R$ MIMO system, the input-output relationship must be instantiated between each pair of transmit and receive antennas. An illustrated summary of OFDM-based MIMO transmission is provided in *Figure 7.3*. In this figure $\mathbf{t}_{mT} = [t_{mT}[1], \ldots, t_{mT}[N]]^T$ is the frequency-domain OFDM symbol transmitted from the m_T^{th} transmit antenna. Analogously,

$\mathbf{y}_{mR}=[y_{mR}[1],\ldots,y_{mR}[N]]^{T}$ is the frequency-domain received vector at the m_R^{th} receive antenna. The N-dimensional receiver noise vector on the m_R^{th} receive antenna is denoted as $\mathbf{n}_{mR}=[n_{mR}[1],\ldots,n_{mR}[N]]^{T}$.

Clearly, OFDM converts the frequency-selective MIMO channel into a set of N orthogonal flat-fading MIMO channels, over which ISI-free transmission can be deployed. Consequently, per subcarrier data detection is sufficient. More importantly, OFDM enables N orthogonal narrowband MIMO transmissions, which can be modulated and detected separately. Consequently, the MIMO-OFDM system model can, and will, in the remaining of this chapter, be simply described per subcarrier:

$$
\begin{bmatrix} y_1[n] \\ \vdots \\ y_{M_R}[n] \end{bmatrix} = \mathbf{H}[n] \cdot \begin{bmatrix} t_1[n] \\ \vdots \\ t_{M_T}[n] \end{bmatrix} + \begin{bmatrix} n_1[n] \\ \vdots \\ n_{M_R}[n] \end{bmatrix} \tag{7.1}
$$

where $\mathbf{H}[n]$ is the M_RxM_T flat fading MIMO channel matrix on the nth subcarrier, defined as follows:

$$
\mathbf{H}[n] = \begin{bmatrix} h_{1,1}[n] & \cdots & h_{1,M_T}[n] \\ \vdots & \ddots & \vdots \\ h_{M_R,1}[n] & \cdots & h_{M_R,M_T}[n] \end{bmatrix} \tag{7.2}
$$

where $h_{i,j}[n]$ is simply the frequency response of the SISO channel between the jth transmit and the ith receive antenna, on the nth OFDM subcarrier.

7.4.2 Overview of MIMO-OFDM processing techniques

Consider the generic MIMO-OFDM communication system in *Figure 7.3*. On the nth subcarrier, a QAM-modulated symbol stream is fed into a MIMO pre-processing block performing the functions of space-time weighting and mapping on M_T parallel streams. These streams, which may range from independent to partially redundant, are mapped on the M_T transmit antennas. At the receiver, on subcarrier n, the M_R receive streams are channelled into a MIMO post-processing block that carries out MIMO channel equalization and demapping. Nevertheless, different MIMO pre-processing and post-processing strategies strike different diversity-multiplexing gain-array gain trade-offs, depending on their level of intelligence, complexity and a-priori channel knowledge.

In this section, we provide a brief overview of the two main trends in MIMO processing, namely the diversity-maximizing space-time coding and the data-rate maximizing spatial multiplexing. The goal is to review their prominent state-of-the-art instantiations, and through them illustrate the key benefits of MIMO communications. For a more exhaustive overview, interested readers are referred to [11][9] and the references therein.

7.4.2.1 Space-time coding

Space-time coding (STC) refers to the set of schemes aimed at the joint encoding of data across both time and transmit antennas. Through appropriate space-time pre- and post-processing, these schemes maximize the spatial diversity gain and/or the coding gain. The first attempt to develop STC was presented in [12] and was inspired by the delay-diversity scheme of Wittneben [13]. STC, however, was truly revealed in [4], in the form of trellis codes. Because trellis codes require complex multi-dimensional vector Viterbi decoding, the popularity of STC really took off with the discovery of space-time block codes [5][6]. The latter codes require only simple linear decoding to achieve the same diversity gain as trellis codes. In the sequel, we briefly review the basic concepts of both space-time trellis and block codes for MIMO systems.

7.4.2.1.1 Space-Time Trellis Coding (STTC)

For every input data symbol $s[k]$, the space-time encoder generates an M_T-dimensional spatial code vector $\mathbf{c}[k]=[c_1[k],...,c_{MT}[k]]^T$, to be transmitted through the M_T transmit antennas. In the framework of space-time trellis coding, an N_F-symbol data frame should be considered. Correspondingly, the transmitted code vector sequence, $\mathbf{C}$, is given by:

$$\mathbf{C} = \left\{ \mathbf{c}[k], ..., \mathbf{c}[k + N_F + 1] \right\} \tag{7.3}$$

which is defined by the underlying trellis description of the code. *Figure 7.4* provides such a trellis description for an 8-PSK eight-state STTC designed for 2 transmit antennas.

On the row of a given trellis state, the ith edge label c_1c_2 indicates that code c_1 is transmitted over the first transmit antenna and that code c_2 is transmitted over the second transmit antenna, for transition to the ith trellis state from the current state. This rate-one code has a bandwidth efficiency of 3 bits per channel use. To determine the performance of STTC, we consider the probability that the decoder decides erroneously in favour of another legitimate code vector sequence, $\hat{\mathbf{C}}$. The resulting $M_T \times M_T$ error covariance matrix $\mathbf{E}$ is expressed as:

$$E\left(C,\hat{C}\right)= \sum_{j=0}^{N_F-1}\left(c[k + j]- \hat{c}[k + j]\right)\left(c[k + j]- \hat{c}[k + j]\right)^* \qquad (7.4)$$

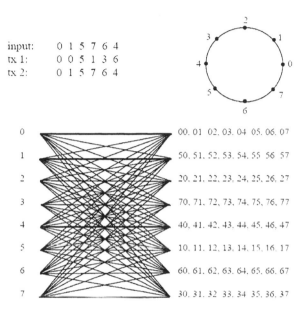

input:	0 1 5 7 6 4
tx 1:	0 0 5 1 3 6
tx 2:	0 1 5 7 6 4

0	00, 01 02, 03, 04 05, 06, 07
1	50, 51, 52, 53, 54, 55 56 57
2	20, 21, 22, 23, 24, 25, 26, 27
3	70, 71, 72, 73, 74, 75, 76, 77
4	40, 41, 42, 43, 44, 45, 46, 47
5	10, 11, 12, 13, 14, 15, 16, 17
6	60, 61, 62, 63, 64, 65, 66, 67
7	30, 31, 32 33, 34 35, 36, 37

Figure 7.4. Trellis description of an 8-PSK 8-state STTC with 2 transmit antennas.

Given perfect channel state information and multi-dimensional vector Viterbi decoding, the pair-wise error probability, i.e. the probability of transmitting **C** and deciding in favour of $\hat{\mathbf{C}}$, is upper bounded for Rayleigh fading channels by [11]:

$$P(\mathbf{C} \rightarrow \hat{\mathbf{C}}) \le \left(\prod_{l=1}^{r}\beta_l^{1/r}\right)^{-rM_R} \cdot \left(\frac{E_s}{4\sigma_n^2}\right)^{-rM_R} \qquad (7.5)$$

where E_s is the symbol energy, r and $\{\beta_l\}_{1\le l\le r}$ are the rank and the non-zero eigenvalues of the error covariance matrix **E**, respectively. Consequently, the STTC performance can then be characterized by

- the achieved coding gain, $g_r = \prod_{l=1}^{r}\beta_l^{1/r}$.

- the extracted spatial diversity order, $r.M_R$ defined as the exponent of the factor $(E_s/4\sigma_n^2)^{-1}$ in (7.5). Since $r\le M_T$, STTC have the capability to achieve the full MIMO spatial diversity order of $M_T M_R$.

Clearly, the STTC should be designed such that both the rank of the error covariance matrix, r, and the coding gain, g_r, are maximized. In this perspective, there has been extensive research to improve on the original hand-crafted STTC designs by Tarokh et al. [4], and to systematize the STTC codes construction. A summary of these research efforts can be found in [14][11].

7.4.2.1.2 Space-Time Block Coding (STBC)

In trying to overcome the STTC decoding complexity, which grows exponentially as a function of the diversity order and the constellation size, Alamouti came up with a now-notorious space-time block coding scheme for two transmit antennas [5]. This remarkable scheme provides full MIMO spatial diversity using simple linear reception. Although it was generalized for arbitrary number of transmit antennas [6], the original Alamouti scheme remains the most powerful and most popular space-time code. For illustration, we herein review its operation for a 2×2 MIMO system, which is also depicted in *Figure 7.5*.

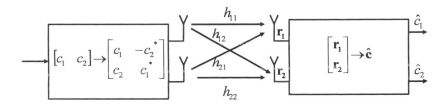

Figure 7.5. Illustration of 2 x 2 Alamouti space-time block coding.

The input symbols to the space-time block encoder are divided into groups of two symbols each, $\{c_1, c_2\}$. Each group is coded and transmitted across the 2 transmit antennas, during 2 consecutive symbol periods, as follows. During the first symbol period, c_1 is transmitted from transmit antenna 1, while c_2 is transmitted from transmit antenna 2. During the next symbol period, $-c_2^*$ is transmitted from transmit antenna 1 and c_1^* is transmitted from transmit antenna 2. Defining $\mathbf{c} = [c_1, c_2]^T$ as the transmit code vector, and $\mathbf{r}_i = [r_i[k], r_i^*[k+1]]^T$ as the receive signal vector at the ith receive antenna during the 2 symbol periods, the following system input-output relationship can be drawn:

$$\begin{bmatrix} \mathbf{r}_1 \\ \mathbf{r}_2 \end{bmatrix} = \begin{bmatrix} \mathbf{H}_1 \\ \mathbf{H}_2 \end{bmatrix} \cdot \mathbf{c} + \begin{bmatrix} \mathbf{n}_1 \\ \mathbf{n}_2 \end{bmatrix} \tag{7.6}$$

where the receiver noise vector at the ith receive antenna is correspondingly defined as $\mathbf{n}_i = [n_i[k], \; n_i^*[k+1]]^T$, and each channel matrix, $\{\mathbf{H}_i, i=1,2\}$, is defined as:

$$\mathbf{H}_i = \begin{bmatrix} h_{i,1} & h_{i,2} \\ h_{i,2}^* & -h_{i,1}^* \end{bmatrix} \tag{7.7}$$

where $h_{i,j}$ denotes the scalar channel between the ith receive antenna and the jth transmit antenna. The major assumption here is that the channel remains constant over two consecutive symbol periods. Realizing that each channel matrix $\mathbf{H}_i$ is orthogonal regardless of the actual channel coefficients, and that $\mathbf{H}_i^H \mathbf{H}_i = (|h_{i,1}|^2 + |h_{i,2}|^2)\mathbf{I}$, it is straightforward to see that simple linear reception is sufficient to un-mix the transmit data symbols at the receiver:

$$\begin{bmatrix} \hat{c}_1 \\ \hat{c}_2 \end{bmatrix} = \begin{bmatrix} \mathbf{H}_1^H & \mathbf{H}_2^H \end{bmatrix} \begin{bmatrix} \mathbf{r}_1 \\ \mathbf{r}_2 \end{bmatrix} \Bigg/ \left(\sum_{1 \le i,j \le 2} |h_{i,j}|^2 \right) . \tag{7.8}$$

Furthermore, it is easy to verify that the receive SNR for both transmit data symbols c_1 and c_2 is

$$\mathrm{SNR} = \sum_{1 \le i,j \le 2} |h_{i,j}|^2 \cdot \frac{E_s}{\sigma_n^2}, \tag{7.9}$$

where E_s is the average symbol energy and σ_n the variance of the noise. Clearly, the clever construction of this space-time block code has allowed to extract the full fourth-order spatial diversity, exhibited by the considered 2×2 MIMO system, using simple linear processing at the receiver. Nevertheless, unlike their trellis counterparts, space-time block codes provide no or minimal coding gain. Finally, as aforementioned, the herein reviewed STBC was extended to more than two transmit antennas [6]. In fact, a general technique for constructing STBCs, which provide the maximum diversity of $M_T M_R$ with simple linear reception, was proposed. Moreover, it was shown that rate-one STBCs can be constructed, provided that real signal constellations are used. For general complex constellations, however, it is unknown whether a rate-one STBC with linear reception exists, when $M_T>2$. Again, complementary information can be found in [14][11] and the references therein.

While many results are available for single-user space-time coding, far less is known about its multi-user extensions. In the context of multi-user scenarios, the uplink is usually called the multiple access channel (MACh), in which several multi-antenna subscriber stations are communicating a different information message to the single- or multi-antenna base station. On the other hand, the downlink is known as the broadcast channel (BCh), in which the multi-antenna base station is communicating a different information message to each of its single- or multi-antenna subscriber stations. Both in the MACh and the BCh, the different subscriber stations aim for maximal spatial diversity and coding gains, while coping with co-channel interference. In the case of the MACh, a specific space-time coding scheme, which combines orthogonal multiple accessing with single-user space-time coding, has been described in [15]. Additionally, the combination of the Alamouti scheme [5] with successive interference cancellation was suggested in [16]. In the case of the BCh, [17] introduced a new concept of space-time coding for the BCh, based on the idea of embedding high-rate space-time codes into low-rate ones. Besides, space-time codes taking into account the presence of co-channel interference have been proposed in [18].

7.4.2.2 Spatial multiplexing

Spatial multiplexing (SM), which was originally proposed by Foschini [19] and Paulraj et al. [20], capitalizes on the multiple SISO subchannels underlying the MIMO channel, to transmit several independent parallel data streams, in an attempt to approach the MIMO capacity[1], or, more pragmatically, boost the data rate. There are several flavours of spatial multiplexing, which deploy different spatial weighting and mapping strategies of the multiple transmit data streams onto the M_T transmit antennas. This subsection summarizes the variants of spatial multiplexing and points out their respective advantages and disadvantages.

7.4.2.2.1 Receive-only processing

Spatial multiplexing with receive-only processing simply transmits independent data streams on different transmit antennas, thus maximizing the average data rate of the MIMO transmission.

---

[1] The Shannon capacity of a channel corresponds to the maximum data rate that can be transmitted over that channel with an arbitrarily small error probability.

In order to perform symbol detection, the receiver must un-do the MIMO channel mixing. Assuming $M_T \leq M_R$, one of the various state-of-the-art equalization strategies can be used:

- The Zero-Forcing (ZF) linear equalizer uses a straight channel matrix inversion, resulting in a poor BER performance when the MIMO channel matrix, $\mathbf{H}[n]$, becomes ill conditioned. Such ill conditioning may arise in certain random fading events, or in presence of a strong LOS component [21].

- The Minimum Mean Square Error (MMSE) linear equalizer exploits the additional knowledge of the noise and interference statistics to improve on the ZF equalizer. However, it only leads to limited gains.

- The optimum decoding strategy is Maximum Likelihood Sequence Estimation (MLSE), where the receiver compares all possible data vectors that could have been transmitted, with what is observed:

$$\hat{t}[n] = \arg \min_{t[n]} \left\| y[n] - \mathbf{H}[n]t[n] \right\|_2 \qquad (7.10)$$

However, MLSE decoding exhibits exorbitant complexity for a large number of antennas and high-order modulations. Some reduced-complexity variants have recently been introduced, such as sphere decoding [22].

- A non-linear decoding strategy that achieves a reasonable complexity-performance trade-off is VBLAST [19][23], also known as nulling and cancelling. It consists of a nulling stage, which estimates a row of $\mathbf{t}[n]$ based on a matrix inversion operation. Then, the cancelling stage subtracts the contribution of the estimated symbol to $\mathbf{y}[n]$. The nulling and cancelling continue successively until all rows of $\mathbf{t}[n]$ have been detected.

- All four aforementioned equalization strategies require the knowledge of the MIMO channel $\mathbf{H}[n]$ at the receiver. This entails training overhead, based on which channel estimation can be carried out. This training overhead can be avoided through using blind equalization, which alternatively exploits higher-order statistics [24][25], subspace methods (see Chapter 3 in [26]) and alphabet information [27]. The price paid for avoiding channel training in blind approaches amounts to some limited loss in BER performance, and, more importantly, to an increased computational complexity. Blind equalization is, herein, mentioned only for completeness. Note that the IEEE 802.16 standard, like most standards, foresees training in its transmission format, such that training-based equalization strategies can be pursued.

When $M_T > M_R$, spatial multiplexing with receive-only processing can only transmit up to M_R independent streams to guarantee symbol recovery. Consequently, subset antenna selection should be used to optimally choose the best subset of M_R out of M_T transmit antennas to transmit over [28][29].

In summary, in its quest to increasing the system's data rate, spatial multiplexing with receive-only processing solely goes after the *spatial multiplexing gain* of MIMO systems. In doing so, it gives up on MIMO pre-processing and its related transmit spatial diversity, resulting in the simplest transmitter design. Nevertheless, this comes at the price of an increased receiver complexity to achieve reasonable BER performance. Consequently, it is mostly suited for uplink transmission, as all the computational complexity is shifted to the base station. In that case, its multi-user extension is straightforward.

7.4.2.2.2 Transmit-only processing

Transmit-only processing is the antipodal spatial multiplexing strategy to receive-only processing. It exclusively concentrates the MIMO processing at the transmitter, leaving to the receiver the simple task of detecting the data symbols:

$$y[n] = \hat{t}[n] = \mathbf{H}[n] \cdot \mathbf{F}[n] \cdot t[n] + n[n] \tag{7.11}$$

where $\mathbf{F}[n]$ denotes the MIMO pre-processor, which clearly must perform channel pre-compensation, in addition to MIMO weighting and mapping. Needless to say that spatial multiplexing with transmit-only processing better fits downlink transmission, such that the subscriber station is spared computationally-demanding MIMO detection.

Just like its receive-only counterpart, transmit-only processing can be implemented following a variety of precoding designs:

- ZF precoding chooses the pre-processing matrix, $\mathbf{F}[n]$, such that the product of this matrix and the channel matrix, $\mathbf{H}[n]$, yields a scaled identity matrix. This amounts to pre-equalizing the MIMO channel, such that the transmitted data symbols of $t[n]$ are received on different receive antennas without multi-stream interference. More specifically, the ZF pre-processing matrix is defined as $\mathbf{F}_{ZF}[n] = pinv(\mathbf{H}[n])/\|\mathbf{H}[n]\|_F$, where the pseudo-inverse of the channel matrix, $pinv(\mathbf{H}[n])$, ensures the perfect separation of the data streams at the receiver, while the normalization by the MIMO channel norm, $\|\mathbf{H}[n]\|_F$, fulfils the transmit power constraint. Since the data streams are received each on a different antenna, no receive spatial diversity can be extracted. Moreover, the perfect channel pre-equalization operation may lead to catastrophic performance in presence of

MIMO channel ill-conditioning, as it leads to large imbalances in the transmit power allocation across data streams.

- To avoid the perfect ZF pre-equalization and the related overly conservative transmit power allocation, MMSE pre-processing can be used [30]. It allows residual multi-stream interference below the receiver noise level. This MMSE pre-processor is simply defined as the transpose of the well-known MMSE detector:

$$F_{MMSE}[n] = \mathbf{H}^H[n] \left(\mathbf{H}^H[n]\mathbf{H}[n] + \frac{\sigma_n^2}{E_s} I_{M_T} \right)^{-1} \qquad (7.12)$$

where σ_n^2 is the power of the temporally and spatially-white complex Gaussian receive noise vector, $\mathbf{n}[n]$, and E_s is the average energy of the constellation, from which the symbol of $\mathbf{t}[n]$ are drawn. MMSE precoding was shown to significantly outperform its ZF counter-part [30].

- The BER performance can be further improved through resorting to non-linear Tomlinson-Harashima (TH) precoding [31][32][30], which essentially implements the transpose of a classical Decision-Feedback Equalizer (DFE). To avoid the increase in required transmit power, induced by the feedback section of the DFE, TH precoding foresees a modulo reduction of the output of the feedback filter to the fundamental interval of the symbol constellation. This modulo operation reduces the required transmit power, without impacting the correct operation of the precoder.

Because it yields independent reception of the transmit data streams on separate receive antennas, spatial multiplexing with transmit-only processing relinquishes receive spatial diversity, except if receive antenna subset selection is used [29][33], in the case that the number of spatially-multiplexed data streams is lower than min(M_T,M_R). More importantly, transmit processing additionally requires a-priori knowledge of the channel state, $\mathbf{H}[n]$, which should be acquired either through uplink channel estimation, in Time-Division Duplexing (TDD) systems, or through a dedicated feedback link in both TDD and Frequency-Division Duplexing (FDD) systems. In practice, the available transmit channel state information (CSI) may be noisy or outdated. In this case, spatial multiplexing with transmit processing will suffer from unrecoverable performance degradation, due to the receiver's inability to carry out complementary equalization.

All aforementioned transmit-only processing solutions can be equivalently applied in the downlink of an SDMA system, provided that the number of base station antennas M_T is larger than the number of the receive antennas across all SDMA subscriber stations. More details about transmit-only SDMA processing can be found in [34][30].

7.4.2.2.3 Joint transmit and receive processing

A spatial multiplexing strategy that allies transmit precoding to complementary receive equalization is joint transmit and receive processing, also widely known as joint linear precoding and decoding. It is a promising and powerful design approach, which jointly designs the MIMO pre-processing and post-processing blocks, $\{\mathbf{F}[n], \mathbf{G}[n]\}$, based on the available transmit CSI, for optimal yet low-complexity MIMO processing. The corresponding input-output relationship reads:

$$\hat{t}[n] = G[n] \cdot \mathbf{H}[n] \cdot F[n] \cdot t[n] + n[n] \qquad (7.13)$$

where the linear precoder $\mathbf{F}[n]$ performs optimal weighting and spreading of the data streams across the M_T transmit antennas, such that the transmission structure is adapted to the preferred directions of the MIMO channel, and that simple matched linear decoding, $\mathbf{G}[n]$, achieves MLSE performance. There exist several optimization criteria for the linear precoder $\mathbf{F}[n]$, including minimizing the mean square error [35][36], maximizing the minimum distance between two received data vectors [37], maximizing the minimum signal-to-noise ratio (SNR) [29][38][39], and maximizing the mutual information [40][28][36].

Spatial multiplexing with joint transmit and receive processing exhibits many advantages, compared to its receive-only and transmit-only processing counterparts:

- because it deploys both spatial precoding and decoding, joint transmit and receive processing is able to extract both transmit and receive spatial diversity.
- because both the precoder and decoder are channel-dependent, they allow coherent combining of the MIMO channel energy. In other words, joint linear precoding and decoding is able to exploit both transmit and receive array gain.
- as aforementioned, both receive-only and transmit-only processing suffer serious performance degradation in ill-conditioned channels. This is because, at the transmitter, receive-only processing simply ignores the MIMO channel, while transmit-only processing seeks to suppress it, through pre-equalization. Joint transmit and receive processing alternatively adapts its transmission to the preferred directions of the MIMO channel. Thus, it is inherently resilient against channel ill-conditioning.

- it is scalable to any number of transmit and receive antennas, unlike transmit-only and receive-only processing, which have dimensionality constraints to guarantee symbol recovery.
- because it adapts the transmission to the structure of the channel, simple linear decoding is sufficient to achieve MLSE performance.

Unlike its receive-only and transmit-only counterparts, the joint transmit and receive processing strategy has the capability to readily extract all the benefits of MIMO, with simple linear processing. However, its performance critically depends on the quality of the available transmit CSI. This has motivated the investigation of various practical joint transmit and receive processing designs, which can actually fulfil the promises of this strategy, under a variety of transmit CSI scenarios.

Over the past few years, several multi-user joint transmit and receive optimization solutions have been introduced. They either advocate the orthogonalization of the transmissions for the different SDMA users [41][42], or propose joint designs of the transmitter and the receivers [43][44]. In the latter case, however, no closed-form solutions have been obtained so far.

7.4.3 Obtaining transmitter channel state information

In the context of BFWA networks, channel estimation is carried out at the receiver using the preamble pre-pended to the data payload. Consequently, it is reasonable to assume that quasi-perfect CSI can be made available at the receiver. Knowledge of the channel state, however, is not naturally available prior to transmission and should be purposefully acquired. To solve the transmit CSI acquisition problem, two fundamentally different approaches can be pursued: the open-loop approach based on channel reciprocity and the closed-loop approach based on feedback. This subsection considers both channel acquisition approaches, highlights the specificities of the imperfect channel acquired at the transmitter, and provides a brief overview of the most prominent designs, in each case.

7.4.3.1 Exploiting reciprocity

The open-loop approach to acquire transmit CSI, capitalizes on the fact that, in TDD systems, uplink and downlink communications are multiplexed on the same physical propagation channel at different time slots, to exploit the CSI estimated in one communication direction for the calculation of the precoder employed in the other communication direction. Clearly, the underlying fundamental assumption of this approach is the reciprocity of the channel undergone by the baseband signals in the uplink and downlink directions. Although the propagation channel is inherently reciprocal, the multi-antenna radio frequency transceivers are certainly not. Thus, the open-loop approach is impaired by amplitude and phase mismatches between the multi-antenna radio frequency transceiver branches, which destroy the channel reciprocity between uplink and downlink. To restore this crucial channel reciprocity, the open-loop approach to acquire transmit CSI, requires the calibration of the used multi-antenna front-ends, either via additional calibration circuitry [45] or via over-the-air calibration procedures [46][47].

In addition to multi-antenna front-end mismatches, channel variations can similarly compromise the availability of perfect timely CSI at the transmitter. Indeed, there will always be a delay between the moment a channel realization is observed in the uplink and the moment it is actually used by the transmitter in the downlink. Combined with BFWA channel variations, this delay inevitably leads to outdated CSI at the transmitter. If this outdated CSI is mistakenly used in the linear precoder calculation, it would inevitably lead to degradation in the system's BER performance. To cope with this impairment, several robust Bayesian approaches have been proposed that take into account the uncertainty on the true channel due to channel variations, for (i) beamforming for MISO systems [48][49], (ii) space-time coded MIMO systems [50][51] and (iii) linearly precoded MIMO systems [38].

7.4.3.2 Explicit feedback

An alternative approach, which avoids calibration in TDD systems and, which is also applicable for FDD systems, consists in estimating the CSI at the receiver side, quantizing it, and then conveying it back to the transmitter through a feedback channel [52][53][54][55].

Nevertheless, one of the challenges of this so-called closed-loop approach is to come up with practical linear precoding solutions that incur minimal feedback overhead. The straightforward approach is to quantize the MIMO channel directly and convey the quantized channel coefficients back to the transmitter [48][49][56][57]. The transmitter then determines the precoder, based on the fed-back quantized channel. Unfortunately, direct quantization of the channel matrix is not efficient for large numbers of transmit and receive antennas, since high feedback bandwidths are required to minimize the quantization error.

Alternatively, quantization of the precoder allows for a better compression of the feedback overhead, by exploiting the special structure of the precoding matrices, and their limited number of underlying degrees of freedom [54][55][58][59]. Two main design approaches can be distinguished. On the one hand, direct quantization of the limited number of underlying independent parameters of the precoders [55]. On the other hand, the more effective, quantization of the space of precoding matrices into a finite multi-mode precoder codebook [53][59][60][61]. Furthermore, in the context of OFDM-based systems, the latter approach can capitalize on the observed correlation between the precoders on adjacent OFDM subcarriers [62][63], to further reduce the required amount of feedback through precoder interpolation [62][63][38].

Similar to the open loop system, the closed-loop CSI estimation will be affected by channel variations.

7.5 SMART ANTENNAS IN THE IEEE 802.16 STANDARD

After having reviewed the main benefits as well as a taxonomy of smart antenna systems, we discuss in this section their application to the IEEE 802.16 standard. Having recognized the huge potential of smart antenna systems, the 802.16 Broadband Wireless Access Working Group currently foresees three main roads for their inclusion in the standard. Subsection 7.5.1 discusses adaptive antenna systems, which are intended to extend the range or increase the capacity of a base station. Subsection 7.5.2 examines the inclusion of space-time block coding techniques, which are mainly intended to improve the link reliability. Finally, Subsection 7.5.3 investigates the inclusion of spatial multiplexing techniques, which are mainly intended to increase the spectral efficiency.

7.5.1 Adaptive antenna systems

Adaptive antenna systems (AASs), also known as intelligent antenna systems, deploy a multi-element antenna array at the base station site, to increase the capacity or to extend the range of a base station. This is illustrated in *Figure 7.6*. As detailed in Subsection 7.3.1, different antenna processing techniques may be envisaged, depending on the presence of one or multiple co-channel subscriber stations (single-user versus multi-user case). For practical systems, the multi-user case seems to be most relevant.

In the uplink, optimum reception may be applied, to enhance the desired user's signal, while suppressing its co-channel interferers. Alternatively, SDMA with receive-only processing may be employed, to jointly detect the different co-channel user signals. Provided transmit CSI can be reliably acquired at the base station, the two aforementioned techniques can be similarly applied in the downlink. In that case, optimum transmission optimally weighs the single data stream across the multiple transmit antennas, such as to maximize the post-processing Signal-to-Interference-and-Noise-Ratio (SINR) at the subscriber station. Alternatively, SDMA with transmit-only processing jointly transmits the different co-channel user signals, such that every subscriber station receives its own desired user signal, (almost) free of co-channel interference.

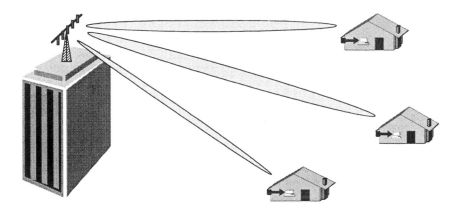

Figure 7.6. Adaptive antenna systems extend range or increase capacity.

The advantages of an AAS are mainly in extending the range and/or increasing the capacity of the base station. On the one hand, by capitalizing on the array gain, diversity gain, and interference rejection gain, optimum combining improves the link budget, which may be translated in a range extension, or, alternatively, a capacity increase through the use of higher modulation orders and coding rates. However, as explained in Chapter 4, the use of an AAS for range extension is a non-trivial task, due to the need for complicated mechanisms to support initial network entry and initial alerting. Nevertheless, in February 2002, Beamreach Networks introduced a BFWA system based on optimum combining, able to support end user service rates ranging from 64 kbps to 1.5 Mbps in the downlink and as high as 1.2 Mbps in the uplink [64]. Their base stations were engineered to handle full link rate and capacity at full cell radius of up to 33 km during NLOS operation. Unfortunately, they gradually ran out of business by the end of 2003. On the other hand, SDMA mainly capitalizes on the spatial multiplexing gain, to offer a significant capacity increase, while only providing limited range extension. In this perspective, ArrayComm's IntelliCell^{TM} fully adaptive smart antenna approach, which includes SDMA, can achieve a spectral efficiency of up to 4 bits/s/Hz/cell, which corresponds to a 40-fold capacity increase compared to classical 3G cellular systems [65].

As explained in Subsection 7.4.3, to acquire transmit CSI reliably, prior to transmission, two basic approaches can be followed. On the one hand, the closed-loop approach, which is suitable for both FDD and TDD systems, estimates and quantizes the CSI at the subscriber station, and then feeds it back to the base station. In this perspective, the standard foresees additional signalling that allows the subscriber station to provide CSI back to the base station. In particular, the standard allows the subscriber station to provide CSI for every 4th, 8th, 16th, 32nd, or 64th subcarrier. On the other hand, the open-loop approach, which is only suitable for TDD systems, relies on the CSI obtained in the uplink direction, to calculate the antenna weights employed in the downlink direction. To enforce the reciprocity between up- and downlink, even in the presence of amplitude and phase mismatches between the multi-antenna RF transceiver branches, additional calibration circuitry or over-the-air calibration procedures are required.

7.5.2 Space-time block coding

A special option of the standard, consists of Alamouti's space-time block coding (STBC) scheme in the downlink, where the base station deploys two transmit antennas. *Figure 7.7* shows a block diagram of the STBC-OFDM transmitter. First, the data is scrambled, coded, and modulated in the traditional way. Next, as detailed in Subsection 7.4.2.1.2, two symbols are taken and alternately transmitted on the two antennas. For each antenna, the OFDM modulation, burst formatting, and radio transmission, are identical to the single-antenna case.

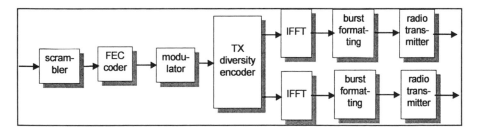

Figure 7.7. Block diagram of STBC-OFDM transmitter.

At the receiver, which is depicted in *Figure 7.8*, two consecutive combinations of the transmitted symbols are received. In the diversity combiner, these original symbols are reconstructed from the two combinations, according to the procedure detailed in Subsection 7.4.2.1.2. To this end, the propagation channel between each transmit antenna and the receive antenna has to be estimated. However, this complexity increase is largely compensated by a second-order diversity gain, which results in a far better performance than single-antenna transmission. Due to its low complexity, space-time block coding may be equally well applied in the uplink, where the subscriber station deploys two transmit antennas, and the base station deploys one or more receive antennas.

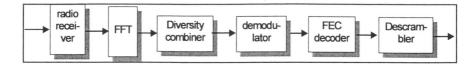

Figure 7.8. Block diagram of STBC-OFDM receiver.

Space time block coding mainly capitalizes on the diversity gain, to significantly enhance the reliability of the link. In specific, Alamouti's space-time block coding scheme is capable to extract the full $2M_R$ diversity order, with M_R the number of receive antennas, by capitalizing on the same underlying space-time block code. This enhanced link reliability leads to a significantly improved receiver performance compared with single-antenna transmissions. Alternatively, the improved link budget can be translated into a capacity increase by resorting to higher modulation formats and coding rates. On the other hand, the standard does not support the translation of the improved link budget into a range extension.

7.5.3 Spatial multiplexing

Spatial multiplexing techniques implicitly refer to a MIMO antenna system, which deploys multiple antennas at both the base station and the subscriber station. As detailed in Subsection 7.4.2.2, different MIMO processing techniques strike different array gain – diversity gain – multiplexing gain tradeoffs, depending on their level of intelligence, complexity, and a-priori channel knowledge. It is important to note, however, that the inclusion of these advanced spatial multiplexing techniques is only foreseen in the revised IEEE 802.16e standard for combined fixed and mobile operation in licensed bands [14].

Spatial multiplexing with receive-only processing is particularly appealing for the uplink, since it concentrates most of the processing burden in the base station, while enabling a very simple transmitter design at the subscriber station. While this processing technique mainly capitalizes on the multiplexing gain, some additional receive array gain and receive diversity gain can be exploited, depending on which type of receiver structure is employed. As such, different receiver structures strike a different trade-off between performance and complexity. On the one hand, linear receiver structures, such as the ZF and MMSE detectors, are easier to implement due to their low computational requirements, but suffer from a mediocre performance. On the other hand, non-linear receiver structures, such as the

ML detector, sphere decoding, or spatial interference cancelling receivers such as the BLAST detector, achieve better performance at the expense of substantially increased computational requirements, especially for the ML detector. In this perspective, Iospan Wireless introduced in September 2002 a 2 x 3 MIMO/OFDM-based BFWA system with receive-only MMSE processing, able to achieve peak data rates of 2*6.8 = 13.6 Mbps (two spatially multiplexed data streams) in a 2 MHz channel bandwidth at distances less than 3.2 km from the base station [67]. In the course of 2003, they were acquired by Intel.

Spatial multiplexing with transmit-only processing is especially suited for the downlink, since all of the computational complexity is shifted to the base station, while leaving the subscriber station the simple task of detecting the data. Like spatial multiplexing with receive-only processing, this processing technique mainly goes after the multiplexing gain. However, some additional transmit array gain and transmit diversity gain may be exploited, depending on which kind of transmitter structure is employed. On the one hand, linear transmitter structures, such as the ZF and MMSE precoders, incur a low computational complexity, but suffer from a mediocre performance. On the other hand, non-linear transmitter structures, such as Tomlinson-Harashima precoding, achieve better performance at the expense of increased computational complexity.

As indicated in the previous paragraphs, spatial multiplexing with receive-only and transmit-only processing mainly pursue spatial multiplexing gain, at the expense of reduced diversity and array gain, due to lack of processing at the transmit and receive side, respectively. This lack of diversity and array gain may lead to poor link-level performance, which may eventually result in a reduced effective throughput, especially for ill-conditioned low-rank MIMO channels and under low SNR operating conditions (which are likely to happen in a realistic BFWA propagation environment). To tackle this problem, spatial multiplexing with joint transmit and receive processing optimally adapts its transmission to the preferred directions of the MIMO channel, resulting in an inherent robustness against channel ill-conditioning. Furthermore, spatial multiplexing with joint transmit and receive processing exploits both transmit and receive array gain as well as transmit and receive spatial diversity gain, which makes it to perform well also under low SNR operating conditions. Due to its acceptable computational requirements, both at the transmit and receive side, spatial multiplexing with joint transmit and receive side is appealing for the uplink as well as the downlink. As already explained in Subsection 7.5.1 for adaptive antenna systems, two basic approaches exist to acquire transmit CSI in a reliable way: the open-loop versus the closed-loop approach, each with its own pros and cons.

7.6 SUMMARY

In this chapter, we have provided a comprehensive overview of smart antenna systems, which are key to meet the data rate and QoS requirements of future broadband wireless services. By additionally exploiting the space dimension, besides the classical time and frequency dimensions, smart antenna systems demonstrate four major benefits. These are the array gain, the diversity gain, the interference rejection gain, and the spatial multiplexing gain. The former three allow to significantly improve the link reliability, and, hence, the QoS and/or the range, while the latter allows to significantly increase the spectral efficiency, and, hence, the supported data rates.

In order to classify smart antenna systems in a structured way, we have introduced a taxonomy, which distinguishes between them based on their system configuration. First, base station antenna systems only deploy multiple antennas at the base station site. Popular antenna processing techniques for this first type of systems, which are applicable to both up- and downlink, are selection diversity, equal gain combining, and maximum ratio combining in the single-user case, and optimum combining and SDMA in the multiuser case. Furthermore, space-time coding techniques are also applicable in the downlink direction.

Second, subscriber station antenna systems refer to systems with a multiple antenna subscriber station and a single antenna base station. Popular antenna processing techniques for this second class of systems are selection diversity, equal gain combining, maximum ratio combining, and optimum combining in the downlink, and space-time block coding, transmit selection diversity, equal gain transmission, and maximum ratio transmission in the uplink.

Third, MIMO antenna systems deploy multiple antennas at both sides of the wireless link. Popular antenna processing techniques for this third and final class of systems are space-time coding techniques, on the one hand, and spatial multiplexing techniques, on the other hand. The former come into two different flavours, namely, high performance space-time trellis coding techniques versus low-complexity space-time block coding techniques. The latter come into three different flavours, that is, spatial multiplexing with receive-only, transmit-only, or joint transmit and receive processing.

In this perspective, MIMO-OFDM is a particularly appealing transmission technique for broadband wireless systems, since it elegantly copes with the frequency- and space-selective fading effects of the underlying broadband MIMO channel. Apart from the sheer accumulation of the respective advantages of its constituent techniques, MIMO-OFDM results in a high implementation efficiency, by exploiting the inherent subcarrier parallelism.

Finally, the IEEE 802.16 Broadband Wireless Access Working Group has recognized the huge potential of smart antenna systems, by providing three parallel tracks for their inclusion in the standard. First, adaptive antenna systems exploit channel state information in both transmit and receive mode, to point a beam in the direction of the desired user, while nulling out unwanted interfering users. While mostly applicable to the base station site, these systems aim at either extending the range (through improving the link reliability) or increasing the capacity of the base station. Second, space-time block coding techniques, which capitalize on the diversity gain, mainly aim at improving the link reliability. Due to their low complexity, these techniques can be readily deployed in the base station as well as the subscriber station. Finally, spatial multiplexing techniques, which mainly capitalize on the spatial multiplexing gain, aim at significantly increasing the spectral efficiency in both up- and downlink directions.

7.7 REFERENCES

[1] K. Sheikh, D. Gesbert, D. Gore, A. J. Paulraj, "Smart Antennas for Broadband Wireless Access Networks", IEEE Communications Magazine, Vol. 37, No. 11, pp. 100-105, November 1999.

[2] D. Gesbert, L. Haumonte, H. Boelcskei, R. Krishnamoorthy, A. J. Paulraj, "Technologies and Performance for Non-Line-of-Sight Broadband Wireless Access Networks", IEEE Communications Magazine, Vol. 40, No. 4, pp. 86-95, April 2002.

[3] R. D. Murch, K. B. Letaief, "Antenna Systems for Broadband Wireless Access", IEEE Communications Magazine, Vol. 40, No. 4, pp. 76-83, April 2002.

[4] V. Tarokh, N. Seshadri, A. R. Calderbank, "Space-time codes for high data rate wireless communication: Performance criterion and code construction", IEEE Transactions on Information Theory, Vol. 44, No. 2, pp. 744-765, March 1998.

[5] S. M. Alamouti, "A simple transmit diversity technique for wireless communications", IEEE Journal on Selected Areas in Communications, Vol. 16, No. 8, pp. 1451-1458, October 1998.

[6] V. Tarokh, H. Jafarkhani, A. R. Calderbank, "Space-time block codes from orthogonal designs", IEEE Transactions on Information Theory, Vol. 45, No. 5, pp. 1456-1467, July 1999.

[7] J. H. Winters, "Optimum combining in digital radio with cochannel interference", IEEE Journal on Selected Areas in Communications, Vol. 2, No. 4, pp. 528-539, July 1984.

[8] R. G. Vaughan, J. B. Andersen, "Antenna diversity in mobile communications", IEEE Transactions on Vehicular Technology, Vol. 36, No. 4, pp. 149-172, November 1987.

[9] A. Paulraj, R. Nabar, D. Gore, "Introduction to space-time wireless communications", Cambridge University Press, May 2003.

[10] R. W. Heath Jr., S. Sandhu, A. Paulraj, "Antenna selection for spatial multiplexing with linear receivers", IEEE Communications Letters, Vol. 5, No. 4, pp. 142-144, April 2001.

[11] D. Gesbert, M. Shafi, D.-S. Shiu, P. J. Smith, A. Naguib, "From Theory to Practice: An overview of MIMO space-time coded wireless systems", IEEE Journal on Selected Areas in Communications, Vol. 21, No. 3, pp. 281-302, April 2003.

[12] N. Seshadri, J. H. Winters, "Two schemes for improving the performance of frequency-division duplex (FDD) transmission systems using transmitter antenna diversity", International Journal of Wireless Information Networks, Vol. 1, pp. 49-60, January 1994.

[13] A. Wittneben, "A new bandwidth efficient transmit antenna modulation diversity scheme for linear digital modulation", IEEE Proceedings of ICC, Vol. 3, pp.1630-1634, May 1993.

[14] A. Naguib, N. Seshadri, R. Calderbank, "Increasing data rate over wireless channels", IEEE Signal Processing Magazine, Vol. 17, pp. 76-92, May 2000.

[15] B. K. Ng, E. S. Sousa, "On bandwidth-efficient multi-user space-time signal design and detection", IEEE Journal on Selected Areas in Communications, Vol. 20, pp. 320-329, February 2002.

[16] A. Stamoulis, N. Al-Dahir, A. R. Calderbank, "Further results on interference cancellation and space-time block codes", Proceedings of Asilomar conference on Signals, Systems and Computers, pp. 257-261, November 2001.

[17] S. N. Diggavi, N. Al-Dahir, A. R. Calderbank, "Diversity embedding in multiple antenna communications", Advances in Network Information Theory, ch. IV, pp. 285-302, American Mathematical Society, DIMACS, 2004.

[18] A. Arunachalam, H. E. Gamal, "Space-time coding for MIMO systems with co-chanel interference", IEEE Transactions on Communications, *to appear*, January 2005.

[19] G. J. Foschini, "Layered space-time architecture for wireless communication in a fading environment when using multiple antennas", Bell Labs Technical Journal, Vol. 1, No. 2, pp. 41-59, September 1996.

[20] A. Paulraj, T. Kailath, "Increasing capacity in wireless broadcast systems using distributed transmission/directional reception (DTDR)", U. S. Patent No. 5345599, Stanford University, September 1994.

[21] D. Gesbert, H. Boelcskei, D. A. Gore, A. J. Paulraj, "Outdoor MIMO wireless channels: models and performance prediction", IEEE Transactions on Communications, Vol. 50, No. 12, pp. 1926-1934, December 2002.

[22] M. O. Damen, A. Chkeif, J. C. Belfiore, "Lattice codes decoder for space-time codes", IEEE Communications Letters, Vol. 4, pp. 161-163, May 2000.

[23] G. D. Golden, G. J. Foschini, R. A. Valenzuela, P. W. Wolniansky, "Detection algorithm and initial laboratory results using V-BLAST space-time communication architecture", Electronics Letters, Vol. 35, No. 1, pp. 14-15, January 1999.

[24] J. Cardoso, A. Souloumiac, "Blind beamforming for non-Gaussian signals", IEEE Proceedings, Vol. 140, part F, pp. 362-370, 1993.

[25] C. Papadias, "A multiuser Kurtosis algorithm for blind source separation Blind beamforming for non-Gaussian signals", IEEE Proceedings of ICASSP, pp. 3144-3147, 2000.

[26] G. Giannakis, H. Hua, P. Stoica, L. Tong, Editors, "Signal processing advances in wireless and mobile communications", Englewood Cliffs, NJ: Prentice-Hall, 2001.

[27] A.-J Van der Veen, A. Paulraj, "An analytical constant modulus algorithm", IEEE Transactions on Signal Processing, Vol. 44, pp. 1136-1155, May 1996.

[28] D. A. Gore, R. U. Nabar, A. Paulraj, "Selecting an optimal set of transmit antennas for a low rank matrix channel", IEEE Proceedings of ICASSP, Vol. 5, pp. 2785-2788, June 2000.

[29] R. W. Heath Jr., S. Sandhu, A. Paulraj, "Antenna selection for spatial multiplexing with linear receivers", IEEE Communications Letters, Vol. 5, No. 4, pp. 142-144, April 2001.

[30] S. Thoen, "Transmit optimization for OFDM/SDMA-based wireless local area networks", PhD dissertation, Katholieke Universiteit Leuven, Belgium, May 2002.

[31] M. Tomlinson, "New automatic equalizer employing modulo arithmethic", IEEE Electronics Letters, Vol. 7, pp. 138-139, March 1971.

[32] H. Harashima, H. Mikayawa, "Matched-transmission technique for channels with intersymbol interference", IEEE Transactions on Communications, Vol. 20, pp. 774-780, August 1972.

[33] H. Sampath, P. Stoica, A.J. Paulraj, "Generalized linear precoder and decoder design for MIMO channels using the weighted MMSE criterion", IEEE Transactions on Communications, Vol. 49, No. 12, pp. 2198-2206, December 2001.

[34] P. Vandenameele, "Space division multiple access for wireless local area networks", PhD dissertation, Katholieke Universiteit Leuven, Belgium, October 2000.

[35] H. Sampath, P. Stoica, A.J. Paulraj, "Generalized linear precoder and decoder design for MIMO channels using the weighted MMSE criterion", IEEE Transactions on Communications, Vol. 49, No. 12, pp. 2198-2206, December 2001.

[36] A. Scaglione, P. Stoica, S. Barbarossa, G. B. Giannakis, H. Sampath, "Optimal designs for space-time linear precoders and decoders", IEEE Transactions on Signal Processing, Vol. 50, No. 5, pp. 1051-1064, May 2002.

[37] L. Collin, O. Berder, P. Rostaing, G.Burel, "Optimal minimum distance-based precoder for MIMO spatial multiplexing systems", IEEE Transactions on Signal Processing, Vol. 52, No. 3, pp. 617-627, March 2004.

[38] N. Khaled, "Joint Transmit and Receive Optimization for MIMO/OFDM-based High-Throughput Wireless Local Area Networks", PhD dissertation, Katholieke Universiteit Leuven, Leuven, Belgium, December 2005.

[39] D. J. Love, R. W. Heath Jr., "Multimode precoding for MIMO wireless systems", IEEE Transactions on Signal Processing, Vol. 53, pp. 3674-3687, October 2005.

[40] G. G. Raleigh, J. M. Cioffi, "Spatio-temporal coding for wireless communications", IEEE Transactions on Communications, Vol. 46, No. 3, pp. 357-366, March 1998.

[41] A. Bourdoux, N. Khaled, "Joint TX-Rx optimization for MIMO-SDMA based on a null-space constraint", IEEE Proceedings of VTC-fall, Vol. 1, pp. 171-174, September 2002.

[42] Q. H. Spencer, A. Lee Swindlchurst, M. Haardt, "Zero-forcing methods for downlink spatial multiplexing in multi-user MIMO channels", IEEE Transactions on Signal Processing, Vol. 52, No. 2, pp. 461-471, February 2004.

[43] K.-K. Wong, R. D Murch, R. S.-K. Cheng, K. B. Letaief, "Optimizing the spectral efficiency of multiuser MIMO smart antenna systems", IEEE Proceedings of WCNC, pp. 426-430, July 2000.

[44] D. Perez Palomar, J. M. Cioffi, M. Angel Lagunas, "Joint Tx-Rx beamforming design for multicarrier MIMO channels: A unified framework for convex optimization", IEEE Transactions on Signal Processing, Vol. 51, No. 9, pp. 2381-2401, September 2003.

[45] J. Liu, A. Bourdoux, J. Craninckx, P. Wambacq, B. Côme, S. Donnay, A. Barel, "OFDM-MIMO WLAN BS Front-end Gain and Phase Mismatch Calibration", IEEE Proceedings of RAWCON, pp. 151-154, September 2004.

[46] J. Ketchum et al., "Qualcomm proposal for MIMO WLAN", IEEE 802.11 document 04/0721r0, July 2004.

[47] S. A. Mujtaba and et al., "TGn Sync proposal technical specification", IEEE 802.11 document 04/0889r44, March 2005.

[48] A. Narula, M. J. Lopez, M. D. Trott, G. W. Wornell, "Efficient use of side information in multiple-antenna data transmission over fading channels", IEEE Journal on Selected Areas in Communications, Vol. 16, No. 8, pp. 1423-1436, October 1998.

[49] E. Visotsky, U. Madhow, "Space-time transmit precoding with limited feedback", IEEE Transactions on Information Theory, Vol. 47, No. 6, pp. 2632-2639, September 2001.

[50] G. Jöngren, M. Skoglund, B. Ottersten, "Combining beamforming and orthogonal space-time block coding", IEEE Transactions on Information Theory, Vol. 48, No. 3, pp. 611-627, March 2002.

[51] S. Zhou, G. B. Giannakis, "Optimal transmitter eigen-beamforming and space-time block coding based on channel mean feedback", IEEE Transactions on Signal Processing, Vol. 50, No. 10, pp. 2599-2613, October 2002.

[52] R. S. Blum, "MIMO with limited feedback of channel state information", IEEE Proceedings of ICASSP, Vol. 4, pp. 89-92, April 2003.

[53] W. Santipach, M. L. Honig, "Asymptotic performance of MIMO wireless channels with limited feedback", IEEE Proceedings of MILCOM, Vol. 1, pp. 141-146, October 2003.

[54] D. J. Love, R. W. Heath Jr., "Limited feedback precoding for spatial multiplexing systems using linear receivers", IEEE Proceedings of MILCOM, Vol. 1, pp. 627-632, October 2003.

[55] J. C. Roh, B. D. Rao, "An efficient feedback method for MIMO systems with slowly time-varying channels", IEEE Proceedings of WCNC, Vol. 2, pp. 760-764, March 2004.

[56] G. Jöngren, M. Skoglund, "Utilizing quantized feedback in orthogonal space-time block coding", IEEE Proceedings of GLOBECOM, Vol. 2, pp. 995-999, November 2000.

[57] G. Jöngren, M. Skoglund, "Improving orthogonal space-time block codes by utilizing quantized feedback information", IEEE Proceedings of Symposium on Information Theory, pp. 220, June 2001.

[58] D. J. Love, R. W. Heath Jr., "Limited feedback precoding for spatial multiplexing systems", IEEE Proceedings of GLOBECOM, Vol. 4, pp. 1857-1861, December 2003.

[59] J. C. Roh, B. D. Rao, "MIMO spatial multiplexing systems with limited feedback", IEEE Proceedings of ICC, Vol. 2, pp. 777-782, May 2005.

[60] K. N. Lau, Y. Liu, T. A. Chen, "On the design of MIMO block-fading channels with feedback-link capacity constraint", IEEE Transactions on Signal Processing, Vol. 52, No. 1, pp. 62-70, January 2004.

[61] D. J. Love, R. W. Heath Jr., "Multimode precoding for MIMO wireless systems", IEEE Transactions on Signal Processing, Vol. 53, pp. 3674-3687, October 2005.

[62] J. Choi, R. W. Heath Jr., "Interpolation based transmit beamforming for MIMO-OFDM with limited feedback", IEEE Proceedings of ICC, Vol. 1, pp. 20-24, June 2004.

[63] J. Choi, R. W. Heath Jr., "Interpolation based unitary precoding for MIMO-OFDM with limited feedback", IEEE Proceedings of GLOBECOM, Vol. 4, pp. 214-218, December 2004.

[64] Mason, C.F, "Pushing the wireless envelope", Broadband, Wireless On-Line, Vol. 3, No. 3, available at http://www.shorecliffcommunications.com/magazine/index.asp, March 2002

[65] IntelliCell™ Smart Antenna Solutions, http://www.arraycomm.com

[66] IEEE P802.16e/D8-2005, "Draft Amendment to IEEE Standard for Local and Metropolitan Area Networks – Part 16: Air Interface for Fixed Broadband Wireless Access Systems – Physical and Medium Access Control Layers for Combined Fixed and Mobile Operation in Licensed Bands", May 2005.

[67] H. Sampath, S. Talwar, J. Tellado, V. Erceg, A. Paulraj, "A Fourth-Generation MIMO-OFDM Broadband Wireless System: Design, Performance, and Field Trial Results", IEEE Communications Magazine, Vol. 40, No. 9, pp. 143-149, September 2002.

Chapter 8

Reducing the Total Cost of Ownership
IEEE 802.16 System Issues

Marc Engels

8.1 TOTAL COST OF OWNERSHIP

According to the Wikipedia [1], the total cost of ownership (TCO) is "a type of calculation designed to help consumers and enterprise managers assess direct and indirect costs as well as benefits, related to the purchase of computer hardware (HW) or software (SW)". A TCO ideally offers a final statement reflecting not only the cost of purchase of a certain system, but also all aspects in its further use and maintenance. This includes, for instance, training support for the company personnel and the users of the system. Therefore, TCO is sometimes referred to as "total cost of operation". TCO analysis originated within the Gartner Group in 1987, and has, since then, been further developed into a number of different methodologies and supporting SW tools."

Although TCO was initially proposed for computer HW and SW, it can easily be adopted for products from other sectors or application contexts, like automotive and telecommunications. In the context of a BFWA network, *Figure 8.1* identifies several factors that will contribute to its TCO. For each base station, for instance, not only the equipment and maintenance costs have to be taken into account, but also the cost for renting, or acquiring, the base station location, which is often called the "roof rights". In addition, a backhaul connection with sufficient capacity has to be established towards the base station location. Moreover, we would like to point out that the cost of capital investments and their associated interests, should not be underestimated and have to be included in the TCO. As a consequence, scenarios where these investments can be aligned with the growth in subscribers, also called "pay as you grow", will be favored by the operators.

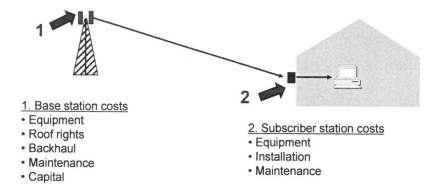

1. Base station costs
• Equipment
• Roof rights
• Backhaul
• Maintenance
• Capital

2. Subscriber station costs
• Equipment
• Installation
• Maintenance

Figure 8.1. Contributions to the TCO of BFWA systems.

The standardized WiMax technology allows for efficient modem implementations that will diminish the equipment cost for the subscriber station. However, also the maintenance costs and, especially, the installation costs are important contributors to the TCO. Ideally, the end-user should be able to install its subscriber station without the need for any in-house wiring.

In this chapter, we focus on several technologies that address the different cost aspects of a BFWA network. More in particular, we will discuss the three different technologies that are shown in *Figure 8.2*.

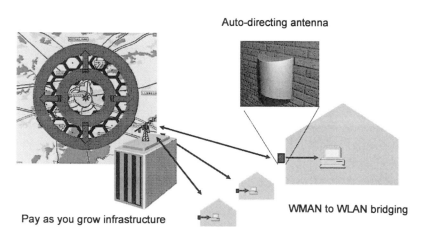

Auto-directing antenna

Pay as you grow infrastructure

WMAN to WLAN bridging

Figure 8.2. Three technologies to reduce the TCO of a BFWA network.

First, auto-directing antennas reduce the installation effort for the customer premises equipment significantly. These antennas automatically detect the direction of the base station and, hence, avoid manual directing of the antenna. Because of their capability of automatically switching to an alternative base station in case of a drop out, they also increase the availability of the BFWA network.

As pointed out in Chapter 3, outdoor subscriber stations can support a larger communication range, but create the additional need for an in-house connection. Seamless integration of Wimax and WiFi technology could provide a wireless end-to-end solution.

Also at the infrastructure side significant improvements can be made with respect to the TCO. Especially, technologies that enable a "pay as you grow" scenario are very attractive, as they reduce the required initial investment of the operators. In the early stages of a deployment, when the subscriber density is low, multi-hop technologies can extend the range of a base station and, hence, reduce the base station costs. The latter not only includes the base station equipment cost, but also roof rights and backhaul costs. When the subscriber density increases, space-division multiple access (SDMA), which was discussed in Chapter 7, can increase the capacity of the base station, hence, postponing the moment when a new base station site with its associated costs needs to be established.

This chapter is organized as follows. First, Section 8.2 introduces the concept of auto-directing antennas. Next, Section 8.3 elaborates on the bridging of wireless local and metropolitan area networks. Section 8.4 subsequently develops a strategy towards "pay as you grow" infrastructure. Finally, Section 8.5 summarizes this chapter with some concluding remarks.

8.2 AUTO-DIRECTING ANTENNAS

The first generation of BFWA customer premises equipment had strongly directive antennas, which extended the range of the equipment, but also complicated the installation process. The equipment had to be installed by skilled personnel of the operator and multiple truck rolls were often necessary. An omni-directional, or isotropic, antenna that has an equal antenna gain in all directions, would simplify the installation but would also reduce the cell size. As a consequence, an adaptive antenna, that automatically directs the antenna beam towards the base station, might be a good option that combines the benefits of both approaches: it has a significant antenna gain and can still be installed by the end user. Moreover, an adaptive antenna can switch over to a back-up base station, in case of loss of connection. As such, it increases the availability of the BFWA network.

An adaptive antenna might be constructed by mechanically moving a directive antenna towards a base station. However, this approach requires a motor, which is costly and subject to failures. Therefore, electronic adaptive antennas are more popular.

The most common way of realizing an electronic adaptive antenna is through a phased array [2]. A phased array antenna combines the signals induced on an array of simple antennas, also called array elements, to form the array output. This process of combining the signals from different elements is also called beamforming. For narrowband signals, the direction of the maximum gain of a phased array antenna can be controlled by adjusting the phase offsets between different antennas by means of phase shifters. To build an adaptive antenna, the phase shifters for the different elements must be programmable, resulting in the block diagram of *Figure 8.3*. With N_a antenna elements, N_a-1 programmable phase shifters are required, resulting in a rather high implementation cost. Programmable phase shifters also introduce a significant RF loss. The adaptive control function monitors the received signal strength, or signal-to-noise ratio (SNR), of the combined signal, and adapts the phase offsets accordingly. Because of the fixed position of the subscriber station, this adaptation can be performed during installation time of the subscriber station, and can be kept fixed from that point onwards.

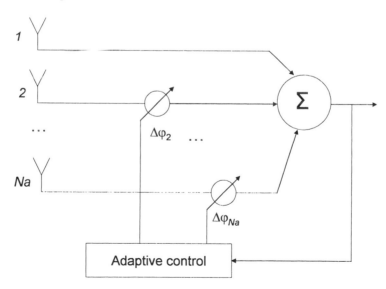

Figure 8.3. Block diagram of an electronic phased array antenna.

Therefore, an alternative approach seems to be more attractive, that is, the switching between N_a directive antenna elements or antenna beams. With a selection device, the antenna element or beam that yields the best performance is selected. An example of this approach for base stations is described in [3]. In [4], the use of a switched directive antenna for BFWA subscriber stations was proposed. The resulting antenna structure can be found in *Figure 8.4*. With N_a antenna elements, N_a single pole, single throw switches are needed.

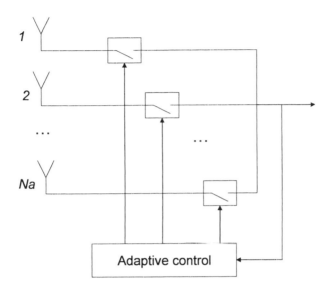

Figure 8.4. Block diagram of an electronic switched antenna.

As a consequence, the complexity of the electronic switched antenna grows with the number of antenna elements, or beams. Also the dimensions of the antenna will increase in the same way. Furthermore, the horizontal antenna beamwidth can be reduced when more antenna elements are used, resulting in a larger antenna gain. However, this increase in directivity is limited by multipath propagation.

Indeed, when the subscriber station is installed at the house of the customer, the signal received by the antenna is very likely to contain a significant portion of multipath radiated components. These multipath components will arrive from different azimuth directions. In [5], this azimuth dispersion, as perceived at a base station for communication in the 1,8 GHz frequency band, is modelled and experimentally validated. For low base station antennas at 20 meters height, roughly equal to the rooftops of the surrounding buildings, the 50% percentile of azimuth spread equals 10°, while the 90% percentile equals 23°. For mobile subscriber stations, it is generally assumed that the received multipath components come from all azimuth directions. However, due to its fixed nature and more elevated position, we expect that the fixed BFWA subscriber stations will still experience a strong directivity of the received signal. Compared to a base station, however, the azimuth spread will be higher. Therefore, it seems reasonable to assume that the 50% percentile azimuth spread, represented by B_{mp}, equals 30°.

To handle this azimuth dispersion at the subscriber stations, the beams of two adjacent antenna elements should overlap at least B_{mp} degrees. Based on this beam overlap and the total pointing range of the adaptive antenna B_{aa}, the 3 dB beamwidth of an antenna element B_{ae} can be calculated as follows:

$$B_{ae} = \frac{B_{aa}}{N_a} + B_{mp} \qquad\qquad (8.1)$$

As a consequence, for a pointing range of 180°, the optimum number of antenna elements ranges between 3 and 6, resulting in 3 dB beamwidths for the antennas between 90° and 60°.

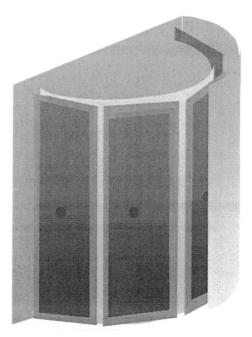

Figure 8.5. Conceptual design of an auto-directing antenna for BFWA.

A conceptual design of such an auto-directing antenna with 4 antenna elements is shown in *Figure 8.5*. The antenna elements are patch antennas, consisting of a ground plane and a two-dimensional radiating structure. The gain in the elevation direction of the antenna is much larger that the azimuthal gain. Therefore, the antennas have an extended vertical dimension.

8.3 WMAN-WLAN BRIDGING

Combining a BFWA subscriber station (SS) with a WLAN access point (AP) in one customer premises equipment might be an attractive concept. Indeed, this combination would eliminate the need for in-house wiring, even with outdoor customer premises equipment. It would also offer the end user the freedom to move around, while staying connected. The concentration of local area and wide area access networking in a single device actually results in a home gateway. Therefore, providing additional services through this gateway is just a small conceptual step. We believe that a business model, in which most of the services are managed and maintained by the operator, is the most attractive one. Also, for WLAN hot spots, BFWA is considered as a good backhaul technology. In this situation, the WLAN hot spot provides high data rates and mobility to several end-users, while the 802.16 technology offers broadband fixed wireless access to the WLAN base station.

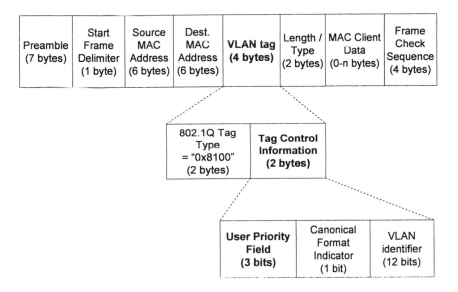

Figure 8.6. The VLAN User Priority Field in an Ethernet frame can be used to communicate the priority of communication streams between BFWA and WLAN.

In both scenarios, home gateways as well as wireless hot spots, the growing importance of real-time multimedia traffic, e.g., voice, music, video, etc., calls for an efficient management of the quality-of-service (QoS) over these wireless links. Service flows with specification of the QoS are an integral part of the IEEE 802.16 standard. Also at the WLAN access point, QoS features will be integrated in the recently finalized IEEE 802.11e standard Remark that QoS support existed already in the ETSI HIPERLAN standard, which did not break through on the market though.

In the European STRIKE project [6], the cooperation in a tandem system of BFWA and WLAN was studied. The interworking between the BFWA subscriber station and the WLAN access point has the following two major functions:

- First, connecting the BFWA service flows to the WLAN QoS classes for the data streams,
- Second, signaling temporary congestion on the WLAN network to the BFWA subscriber station.

The QoS connection of the data streams can be implemented at various levels: data link control (DLC), Ethernet, or IP. The Ethernet bridging is the most easy to integrate with the WLAN standards, and still provides maximum support for higher-layer protocols. In this approach, the extended Ethernet frame format, defined by the IEEE 802.3ac standard for Virtual Local Area Networks (VLANs), is used. As illustrated in *Figure 8.6*, the 3-bit user priority field of the VLAN tag is used to associate a connection at both the WLAN and the BFWA side. The major drawback is the fact that the user priority field only allows for eight different traffic classes.

Congestion control can be performed at the higher-layer end-to-end protocols, or at the data link layer. Basic mechanisms for end-to-end congestion control in TCP networks exist for a long time [7]. However, as pointed out in [8], they are not very well suited for wireless networks. Therefore, the inclusion of congestion control at the data link layer is often advocated, rather than delegating it to the higher end-to-end layers. Congestion control at the data link layer can react faster and adapt better to the link characteristics. Furthermore, it would make congestion control available to all higher-layer protocols.

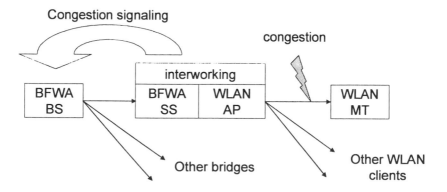

Figure 8.7. Congestion in WLAN downlink.

The European STRIKE project investigated the use of an additional medium access control (MAC) management message in the BFWA network to signal congestion in the WLAN to the BFWA base station. As illustrated in *Figure 8.7* for downlink communication, the MAC management message would signal back congestion in the WLAN downlink, such that the BFWA base station can temporarily reduce the transmitted data rate. In the uplink, which is shown in *Figure 8.8*, the MAC management message is of particular interest in the unsolicited grant service. If the WLAN uplink is congested, the message signals the base station to temporarily reduce the allocated bandwidth for this specific connection. The inclusion of MAC messages for congestion notification is completely in line with the cross-layer congestion control, as advocated in [9].

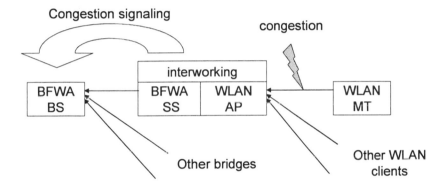

Figure 8.8. Congestion in WLAN uplink with unsolicited grant service.

Data link layer congestion control also requires a mechanism to detect congestions on the WLAN link. Congestion detection can be performed by the WLAN access point, and, subsequently signaled to the BFWA subscriber station, or it can be realized directly by the BFWA subscriber station. The mechanism can be based on monitoring of the buffers for the data streams: overflow detection for the downlink and underflow detection of the unsolicited grant service for the uplink. However, monitoring buffer over- and under-flows can result in false connection triggers, especially for variable bit rate sources. Therefore, communication latencies are a more reliable measure for congestion in wireless networks [10].

To validate the performance of the WMAN-WLAN interworking, simulations were performed for various situations. The schematic set-up for one particular interworking example is shown in *Figure 8.9*. The system consists of a BFWA base station (BS), an interworking gateway that contains a BFWA subscriber station (SS) and a WLAN access point (AP), and two WLAN mobile terminals (MT1 and MT2). MT1 has a low-priority best-effort bidirectional communication with a maximum throughput of 15 Mbps in both directions. After 20 seconds, a high-priority bidirectional communication of 11 Mbps in each direction is started up for MT2, representing a high-quality two-way video stream. Furthermore, the total throughput in the WLAN is limited to 40 Mbps.

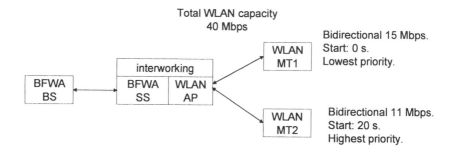

Figure 8.9. Schematic set-up for an interworking example.

Figure 8.10 illustrates the evolving throughput over time for the three different connections. When the second data stream is activated, the effective data rate of MT1 is reduced to 9 Mbps, due to the capacity limit of the WLAN. However, the downlink BFWA connection is not aware of the bottleneck and keeps on sending the full 15 Mbps data. This will result in a buffer overflow in the BFWA SS or WLAN AP. Hence, this data will have to be retransmitted, consuming precious capacity resources of the BFWA network that could have been effectively used for other users of the network. When the interworking mechanism is activated, the downstream BFWA throughput is temporarily degraded, such that only the 9 Mbps that can be accommodated by the WLAN AP are sent.

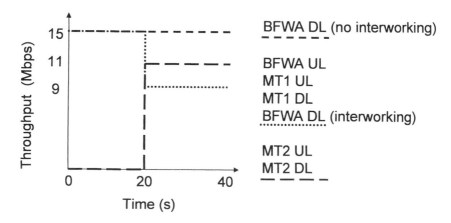

Figure 8.10. Interworking example: throughput over time on the three connections.

The gain of the interworking mechanism for this example is illustrated in *Figure 8.11*. The figure shows the total capacity used for the WLAN and BFWA networks. Initially, both networks transfer 30 Mbps and are effectively used. When the data stream to MT2 becomes active, the throughput on the WLAN is raised to the limit of 40 Mbps. Without interworking, however, the BFWA network transmits 46 Mbps, resulting in a loss of 6 Mbps. The inclusion of the interworking functionality avoids this loss.

Practical experimentation has learned that the adaptation of the data rate of the downlink BFWA traffic takes some time. As a consequence, the dynamic congestion control mechanism is only effective with relatively large update intervals (in the order of seconds). The underlying cause is that the scaling of the data streams is currently still performed by the end-to-end protocol layers. As a consequence, for stream-based applications, end-to-end flow control mechanisms are able to provide an equivalent gain as the data link layer interworking mechanism. The use of scalable data streams, of which the data rate can be adapted to the available capacity in intermediate network nodes, would change this conclusion. Also, short transactional applications should profit more from data link layer congestion control.

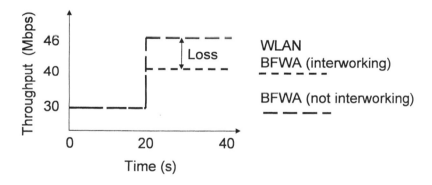

Figure 8.11. Interworking example: total throughputs of the networks.

The data link layer congestion control is significantly more expensive to implement for larger network configurations, as a specific connection needs to be maintained for every scalable data stream. This increased cost might render end-to-end congestion control more attractive.

8.4 PAY AS YOU GROW INFRASTRUCTURE

When an operator starts deploying its network, the subscriber density will be very low. However, for an attractive business proposal, the operator should offer a good coverage of a targeted subscriber area. As a consequence, he has to establish a sufficient amount of base stations. The cost of renting base station sites, also referred to as "roof rights", and operating a backhaul connection, constitutes a considerable investment.

Therefore, the introduction of so-called repeaters that receive, store, and forward the communication to a main base station, might be an attractive proposal. Indeed, these repeaters extend the range of the main base station without the need for extra backhaul connections. Moreover, these repeaters do not need to have premium locations, but could be installed on lamp poles, or subscriber sites. In *Figure 8.12*, such a typical two-step network set-up is compared with a traditional cellular network deployment. For the two-step network architecture with 30 repeaters, the radius of a base station is expanded from 2.7 km to 6.1 km. Therefore, the single base station can cover a region that otherwise would need 7 base station sites.

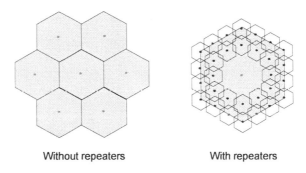

Without repeaters   With repeaters

Figure 8.12. Repeaters increase the coverage area of a base station.

As explained in Chapter 4, a two-step communication will generally halve the capacity of the base station. However, the use of directional antennas, or at least two halve planes, can almost completely eliminate this capacity reduction. The use of sectorized antennas at the base station is common practice. For the subscriber stations as well as the repeaters, the auto-directing antenna that was presented in Section 8.2 can be used.

The key to this optimized capacity is a clever communication scheduler in the base station[11] that optimizes the use of the limited spectrum by allowing several communications between repeaters and subscriber stations and/or between repeaters and base stations to take place in the same time slot. As a consequence, these simultaneous communications might mutually interfere. Therefore, the intelligent scheduler should guarantee that mutually interfering communications are scheduled during different time slots. Moreover, the intelligent scheduler should make sure that each repeater or subscriber station is involved in only one communication in a single time slot.

Directional antennas in the repeaters and the subscriber stations are the key enablers for simultaneous transmissions. For the subscriber station, the crucial requirement is a sufficient front-to-back attenuation of the antenna when it has been installed at the customer premises. Interfering signals that are arriving from the back or from the side have to be attenuated sufficiently in order to allow the demodulation of the desired signal. To allow simultaneous transmissions, a repeater will need two directional antennas: one that is oriented towards the base station (inward), and one that is oriented away from the base station (outward). While the inward antenna communicates with the base station and subscriber stations, the outward antenna solely communicates with subscriber stations. Remark that the two antennas might share a single radio that switches between them, but the use of two radios, one for each antenna, is more straight forward, and will be assumed in the remainder of this section.

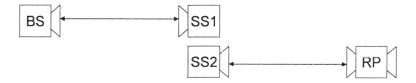

Figure 8.13. Some communications between the base station and subscriber station 1 can take place simultaneously with a communication between the repeater and subscriber station 2.

A basic example of simultaneous communications based on the front-to-back isolation of the subscriber stations is shown in *Figure 8.13*. Two subscriber stations (SS1 and SS2) are positioned close to each other. The base station (BS) communicates with SS1, while the inward antenna of the repeater (RP) communicates with SS2. We assume that the BS, RP, SS1, and SS2 all transmit with the same transmit power, and that the path losses between BS, or RP, and ST1, or ST2, are equal. Under these assumptions, the signal-to-interference ratio (SIR) for a simultaneous downlink communication from the BS to SS1 and from the RP to ST2 equals the front-to-back isolation of the subscriber station antenna. As a consequence, with a sufficient front-to-back isolation these two communications can take place in the same time slot. A similar reasoning can be made for the uplink communications from SS1 to the BS, and from SS2 to the RP.

With simultaneous transmissions taking place, the scheduling of the transmissions between the different elements of the system is crucial for the

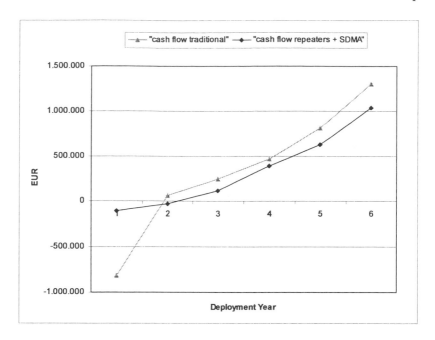

Figure 8.18. The required start-up capital for a "pay as you grow" infrastructure, including SDMA and repeaters, is largely reduced.

8.5 SUMMARY

In this chapter, we have pointed out that lowering the cost of ownership of a BFWA network is a crucial factor for its success. Many aspects of the BFWA solution contribute to this reduction. In this chapter, we have elaborated three examples:

- Auto-directing antennas will simplify the installation of BFWA customer premises equipment, and reduce the number of truck rolls for the operator.
- WMAN-WLAN bridging enables wireless home gateways and hot-spots with wireless feeds. With the growing importance of multimedia applications, QoS provisioning is crucial for this interworking.
- "Pay as you grow" infrastructure combines multi-hop communication and advanced antenna concepts to minimize the required start-up capital for deploying and evolving a BFWA network.

We believe that the above technologies are essential for a widespread adoption of the BFWA technology. However, creative minds can come up with many more cost saving and business generating technologies that will further stimulate the take up of BFWA.

8.6 REFERENCES

[1] Wikipedia, http://www.wikipedia.org/
[2] L. C. Godara, "Applications of Antenna Arrays to Mobile Communications, Part I: Performance Improvement, Feasibility, and System Considerations", Proceedings of the IEEE, Vol. 85, No. 7, pp. 1031-1060, July 1997.
[3] P.C.F. Eggers, B. Krøyer, "Transparent Antenna Solutions for Adaptive Coverage Systems", Wireless Personal Communications, Vol. 30, No. 2-4, pp. 131-152, September 2004.
[4] M. Engels, J. Erreygers, J. Haspeslagh, F. Op 't Eynde, "Device and Method for Improving Wireless Outdoor-to-Indoor Digital Communication", Patent Application WO03058850.
[5] K. I. Pedersen, P. E. Mogensen, B. H. Fleury, "A Stochastic Model of the Temporal and Azimuthal Dispersion Seen at the Base Station in Outdoor Propagation Environments", IEEE Transactions on Vehicular Technology, Vol. 49, No. 2, pp. 437-447, March 2000.
[6] "Spectrally Efficient Fixed Wireless Network based on Dual Standards", www.ist-strike.org
[7] S. H. Low, F. Paganini, J. C. Doyle, "Internet Congestion Control", IEEE Control Systems Magazine, Vol. 22, No. 1, pp. 28-43, February 2002.
[8] G. Xylomenos, G. C. Polyzos, P. Mähönen, Mika Saaranen, "TCP Performance Issues over Wireless Links", IEEE Communications Magazine, Vol. 39, No. 4, pp. 52-58, April 2001.
[9] S. Shakkottai, T.S. Rappaport, P.C. Karlsson, "Cross-Layer Design for Wireless Networks", IEEE Communications Magazine, Vol. 41, No. 10, pp. 74-80, October 2003.
[10] P. Rosson, H. Cappelle, F. Platbrood, P. Wachsmann, "Deliverable D6.7.1: Report on Demonstration Results", IST STRIKE project, September 2004.
[11] M. Engels, J. Erreygers, F. Op 't Eynde, "Wireless Cellular Network Architecture", International Patent, WO 03/058984, January 2003.
[12] P. Vandenameele, L. Van der Perre, M. Engels, "Space Division Multiple Access for Wireless Local Area Networks", Kluwer Academic Publishers, 2001.

Chapter 9

Looking Backward and Forward
Conclusions and Challenges

Marc Engels, Frederik Petré

9.1 CONCLUSIONS

As pointed out in the introductory chapter of this book, three conditions are key to the success of broadband fixed wireless access: standardization, high performance, and low total cost of ownership. In the subsequent chapters, we have illustrated that the WiMax technology is rapidly realizing these conditions.

9.1.1 A standardized solution

Standardization is an essential factor to come up with interoperable equipment. Chapter 2 has showed that multiple technologies compete to provide broadband wireless access: IEEE 802.11, IMT-2000, IEEE 802.20, and IEEE 802.16. For fixed applications, the emerging IEEE 802.16 standard, especially operating below 11 GHz, has been identified as the most promising technology. Therefore, we have restricted our further discussions to this standard.

Before elaborating on the technical details of the IEEE 802.16 standard, we have reviewed in Chapter 3 the characteristics and the main impairments of the BFWA radio channel. For the outdoor path loss of a BFWA system we have introduced a statistical model that is derived from a large measurement campaign and that is adopted by the standardization committee. In addition, for a CPE that is located indoor, the outdoor-to-indoor penetration loss had to be taken into account, with practical values well above 15 dB. In addition to path loss, the BFWA radio channel features considerable multipath propagation in combination with limited time variance, causing signal dispersion along three dimensions: the delay, the Doppler frequency, and the angle dimension. Delay dispersion is quantified by the RMS delay spread with typical values for the BFWA radio channel ranging between 0.111 µs and 5.240 µs, depending on the terrain conditions. Doppler shift dispersion induces temporal variations in the channel, which are quantified by the channel's coherence time. As derived from an extensive measurement campaign, typical mean values of the 70 % coherence time vary between 44 ms and 90 ms. Angle dispersion, which is especially relevant to multiple antenna systems, is quantified by the angle spread. For BFWA channels the 50% percentile of the angle spread is expected to be around 10°.

A detailed description of the IEEE 802.16 standard can be found in the two subsequent chapters. In Chapter 4, we have introduced the data link layer, consisting of the MAC and the security functionalities. On top of that, the service-specific convergence sublayers, which form the interface between the higher layer protocols and the data link layer, are pointed out to play a crucial role in the provisioning of QoS features.

In Chapter 5, we have provided a bird's eye view of the different physical layers for broadband fixed wireless access, as described in the IEEE 802.16 standard. We focussed on the three non-compatible PHYs that have been developed for communication in the 2 to 11 GHz frequency bands, based on respectively single carrier, OFDM and OFDMA modulation. Because of the active backing of the WiMAX Forum and major market players, it is believed that the OFDM scheme will become dominant in the BFWA market. In January 2006, the WiMAX forum announced the certification of the first six products [1].

However, standardization work is never fully completed. For instance, the mobile extension (IEEE 802.16e), which supports handover between base stations, was recently approved. Also management information bases and management procedures are currently being standardized. Committees are also studying the license-free operation and mobile mesh networking capabilities. And more might be expected to come!

9.1.2 A high-performance solution

The basic IEEE 802.16 standard, as elaborated in Chapters 4 and 5, already features a high throughput. For instance, in a 14 MHz band, a maximum raw PHY data rate of 44 Mbps, or a bandwidth efficiency exceeding 3 bits/Hz, was reported. Because of the centralized scheduling approach, a high MAC efficiency is achieved: with less than 10 active users per MAC frame, it exceeds 80%. The MAC also guarantees the QoS of the various data streams.

However, further performance improvements can be achieved with new smart antenna techniques. In Chapter 7, we have provided a comprehensive overview of smart antenna systems. These systems allow a significant increase of the spectral efficiency and a significant improvement of the link reliability compared to traditional single antenna systems. As illustrated in this chapter, multiple antennas can be deployed at the base station site or subscriber station, or both. The latter technique is called MIMO processing and can be efficiently combined with OFDM, by exploiting the inherent subcarrier parallelism. Three variants of smart antenna systems are supported by the IEEE 802.16 standard. First, adaptive antenna systems exploit channel state information in both transmit and receive mode, to point a beam in the direction of the desired user, while nulling out unwanted interfering users. Second, low complexity space-time block coding techniques aim at improving the link reliability. Third, spatial multiplexing techniques aim at significantly increasing the spectral efficiency in both up- and downlink.

9.1.3 A solution with a low total cost of ownership

Aggressive integration of 802.16 modems will result in low cost terminals. In Chapter 6, we have reviewed the challenges in designing cost-efficient BFWA transceivers. The overall architecture was our first point of attention. Flexible interfaces between hardware and software and between digital and analog processing were pointed out as crucial features. For the baseband receiver, we have demonstrated a large similarity between frequency-domain processing for single-carrier modulation and for OFDM. More specifically, similar techniques can be used for channel estimation, symbol timing synchronization, and carrier frequency synchronization, for these block-based communication schemes. For the radio front-end, we have advocated the approach of adopting a low-cost direct conversion (or zero IF) radio and the compensation of the resulting analog impairments in the digital domain. Recently, efficient 802.16 modem implementations have been announced by Fujitsu [2], Intel [3], and Wavesat [4]. Also accompanying radio chips are available from several companies.

However, as we have pointed out, more is needed for a successful BFWA business proposition. Additional technologies will have to be deployed to lower the total cost of ownership of BFWA networks for operators. In Chapter 8, we have elaborated on three appealing example technologies:

- First, auto-directing antennas will simplify the installation of BFWA customer premises equipment, and reduce the number of truck rolls for the operator.
- Second, WMAN-WLAN bridging enables wireless home gateways and hot-spots with wireless feeds. With the growing importance of multimedia applications, QoS provisioning is crucial for this interworking.
- Third, "pay as you grow" infrastructure combines multi-hop communication and smart antenna concepts to minimize the required start-up capital for deploying and evolving a BFWA network.

We believe that the above technologies are essential for a widespread adoption of the BFWA technology. However, creative minds can come up with many more cost saving and business generating technologies that will further stimulate the take up of BFWA.

9.2 CHALLENGES

In this book we have explored the fascinating broadband fixed wireless access technology, which seems on the verge of a market breakthrough. The first priority for making WiMax successful is the creation of a supporting business environment. This requires actions from equipment manufacturers, regulatory bodies, network operators, as well as an economic climate that is positive to investments. These aspects are further discussed in Subsection 9.2.1.

Although WiMax equipment is appearing on the market and is expected to generate considerable revenues, the maximum market penetration will not be realized in a short time. This leaves significant room for further research and innovations. Especially, aspects as smart antennas, multi-hop networking, and link adaptation are interesting areas for future research. Some examples are given in Subsection 9.2.2.

It can be expected that WiMax technology will be an essential component of the 4th generation broadband wireless system (4G). However, the 4G system will consists of a horizontal integration of multiple communication technologies on a common IP-based platform. As such cost-efficient and flexible multi-mode terminals will be crucial for its success, as argued in Subsection 9.2.3.

9.2.1 Creating the business environment

As argued in Chapter 1, broadband fixed wireless access technology is ready for take-off. This is not only due to progress in technology, but also due to various business evolutions. These include the following:

- The standardized technology yields the promise of interoperable equipment. However, in order to rapidly converge on a worldwide standard, a staggering number of PHY and MAC options are provided in the IEEE 802.16 standard. Therefore, the WiMax Forum plays an essential role to ensure interoperability of the IEEE 802.16 vendor products. The WiMax Forum defined profiles, which specify, among others, the frequency band of operation, the bandwidth, and the physical layer mode. It offers certifications programs to verify adherence to these profiles.

- The standard has convinced several major chip manufacturers to develop integrated solutions for IEEE 802.16. In combination with the increased competition, these devices will drive the cost of the equipment down.

- The equipment manufacturers have learned from previous experience and offer more robust products that also work under NLOS conditions.

- The availability of spectrum has increased. For instance, in Europe, the band between 3,4 GHz and 3,8 GHz has been reserved for point-to-multipoint fixed access with a harmonized frequency plan [5].

- With the growing interest in high-speed internet and triple play (video, data and voice) offerings, the demand for bandwidth is ever rising, even in developing countries where no extensive wired infrastructure exists.

- The investment climate in the telecom sector has gradually improved over the last years. In addition, several operators are competing fiercely on existing and particularly developing markets.

Based on these positive evolutions, a rapid growth of WiMax equipment revenues is generally expected [6]. However, WiMax technology will not replace DSL or cable modems. It is rather seen as a complementary technology that will be especially successful in the world's rural and developing areas without a broadband fixed infrastructure. With an appropriate business model, possibly combining multiple technologies, WiMax technology might also compete in urban areas for the mobile broadband user [7].

9.2.2 Improving the BFWA technology

Although the standardized 802.16 technology marks a significant step towards broadband wireless access, the continuous demand for higher throughputs and larger coverage areas leaves significant room for further research and innovations. In [8], four potential enhancements were proposed: MIMO processing, Hybrid-ARQ, interference cancellation, and adaptive modulation and coding.

In MIMO techniques, introduced in Chapter 7, multiple antennas are deployed both at the base station and the subscriber station. With these techniques, multiple data streams can be multiplexed or the link reliability can be increased. As such, MIMO allows a trade-off between increased throughput and range. The study of MIMO receiver techniques that strike the right balance between receiver performance and implementation complexity is an active research area. On a longer term, multi-user MIMO, also called MIMO-SDMA [9], techniques can be introduced.

Hybrid-ARQ has already been defined in the 1960's and combines the received data from all retransmissions in an ARQ scheme to improve link reliability. According to [8], it could be applied to IEEE 802.16 with some minor changes in the standard and would increase the data rate in the low SNR regime.

For IEEE 802.16 users that are on the edge of a cell, co-channel interference might be a major problem. As a consequence, there is a growing interest in low-complexity interference cancellation techniques in a subscriber station receiver. For CDMA, this topic has been extensively investigated [10]. In most solutions at least 2 antennas are required at the receiver. Recently, similar studies are being made for OFDM based communication [11].

Similar to IEEE 802.11, also IEEE 802.16 assumes that the modulation on every subcarrier is identical. Nevertheless, it has been demonstrated that adaptive subcarrier modulation would largely improve the performance of a WLAN communication link [12]. The same idea could also be applied to IEEE 802.16 systems and similar improvements can be expected. Moreover, the combination of MIMO with adaptive subcarrier modulation has the potential to even further increase the spectral efficiency and improve the link reliability.

In [13], it was pointed out that the basic mesh networking scheme, as adopted in the 802.16 standard, is not very bandwidth efficient. Therefore, the authors proposed an interference-aware routing and scheduling approach. In general, optimal mesh networking, which takes into account the QoS of the data streams is an open research field.

Also the low-power and flexible implementation of an IEEE 802.16 terminal offers a number of open research challenges. For instance, the IEEE 802.16 standard defines different carrier frequencies, varying channel bandwidths, several duplexing schemes, etc. For mobile users, a flexible terminal that supports the full flexibility of the standard would be useful. By doing so, the mobile terminal would be able to work on the IEEE 802.16 networks of different operators with different settings for these parameters.

The above are just examples of possible enhancements of the IEEE 802.16 technology. Many others could be envisaged and the list is only limited by the imagination of the research community.

9.2.3 Towards 4G

As pointed out in Chapter 2, to realize ubiquitous connectivity, multiple technologies will have to be combined: IMT-2000 or IEEE 802.20 when travelling at high speed, IEEE 802.11 at the office and a combination of IEEE 802.11 and IEEE 802.16 at home or within reach of a public hot spot. As a consequence, multi-standard functionality, where different wireless access systems are combined on a common IP-based platform, is needed, especially at the terminal side. With this aim, the newly established IEEE 802.21 Media Independent Handoff Working Group studies the handoff between these various technologies. The same strategy will be followed for the 4th generation (4G) broadband wireless system [14]. It will employ a horizontal rather than a vertical communication model, which integrates different existing and evolving wireless access systems on a common IP-based platform, to complement each other for different service requirements and radio environments.

To enable seamless and transparent interworking between these different component systems, or communication modes, through horizontal (intra-system) or vertical (inter-system) handovers, multimode terminals are needed that support different existing and evolving air interfaces. The straightforward solution of providing a custom baseband processor for every communication mode and collecting them into a single terminal, would certainly lead to a very costly and power-inefficient terminal implementation. Hence, to enable low-cost multimode terminals with low power consumption and reasonable size, a flexible rather than a fixed air interface concept is needed that not only supports a single access mode but different existing and evolving access modes.

The vertical handover between the different wireless networks is another major challenge for next generation networks. The switching between the various wireless networks is aimed at achieving a maximal performance for a wide variety of application areas and communication environments. Most authors assume that this switching should be user/terminal centric. This requires that the terminal has efficient switching criteria, taking into account the link adaptation of the multiple wireless systems. Moreover, guaranteeing the QoS of the communication during handover is an open research topic that is being addressed by a large research community.

9.3 SUMMARY

This brings us at the end of this book. We have explored the fascinating broadband fixed wireless access technology, which seems on the verge of a market breakthrough. This is enabled by three properties of the technology: standardized, high performance, and low total cost of ownership. Acceptance of the technology can be accelerated by the creation of a supporting business environment and further technological improvements, especially to lower the total cost of ownership. As a consequence, we believe that broadband fixed wireless access technology faces a bright future. We also anticipate that it will become an essential component of the 4th generation broadband wireless system (4G). We therefore hope that this book might serve as a jumpstart for your research or business adventure in broadband fixed wireless access.

9.4 REFERENCES

[1]   http://www.wimaxforum.org/
[2]   "Fujitsu Announces New Highly Integrated WiMAX SoC, Assumes Industry Leadership in IEEE 802.16 Technology", Fujitsu press release, 21 April 2005, http://www.fujitsu.com/us/news/pr/fma_20050421-1.html.
[3]   "Intel Introduces New WiMAX Silicon Solution To Expand The Reach Of Broadband Internet Access", Intel press release, 18 April 2005, http://www.intel.com/pressroom/archive/releases/20050418comp_a.html.
[4]   Wavesat Announces General Availability of Its WiMAX Chip, Wavesat press release, 12 January 2005, http://www.wavesat.com/media/releases/current_year/120105.html.
[5]   "Harmonized Radio Frequency Channel Arrangements and Block Allocations for Low, Medium and High Capacity Systems in the Band 3600 MHz to 4200 MHz", CEPT/ERC/RECOMMENDATION 12-08 E, 1998
[6]   S. J. Vaughan-Nichols, "Achieving Wireless Broadband with WiMax", IEEE Computer, Vol. 37, No. 6, pp. 10-13, June 2004.
[7]   V. Gunasekaran, F.C. Harmantzis, "Migration to 4G-Ubiquitous Broadband Economic Modeling of WiFi with WiMax", Proc. WWC 2005, SFO, USA, May 2005.
[8]   A. Ghosh, D. R. Wolter, J.G. Andrews, R. Chen, "Broadband Wireless Access with WiMax/802.16: Current Performance Benchmarks and Future Potential", IEEE Communications Magazine, February 2005, Vol. 43, No. 2, pp. 129-136.
[9]   P. Vandenameele et al., "Space Division Multiple Access for Wireless Local Area Networks", Kluwer Academic Publishers, 2001.

[10] F. Petré, "Block-Spread CDMA for Broadband Cellular Networks", PhD Thesis, Katholieke Universiteit Leuven, Leuven, Belgium, December 2003.

[11] J. Li, K. B. Letaief, Z. Cao, "Co-Channel Interference Cancellation for Space-Time Coded OFDM Systems", IEEE Transactions on Wireless Communications, Vol. 2, No. 1, pp. 41-49, January 2003.

[12] S. Thoen, "Transmit Optimization for OFDM/SDMA-based Wireless Local Area Networks", PhD Thesis, Katholieke Universiteit Leuven, Leuven, Belgium, May 2002.

[13] H.-Y. Wei, S. Ganguly, R. Izmailov, Z. J. Haas, "Interference-Aware IEEE 802.16 WiMax Mesh Networks", Proceedings of the 61st IEEE Vehicular Technology Conference (VTC 2005 Spring), Stockholm, Sweden, May 29-June 1, 2005.

[14] M. Katz and F. H. P. Fitzek, "On the Definition of the Fourth Generation Wireless Communications Networks: The Challenges Ahead", International Workshop on Convergent Technology (IWCT), Oulu, Finland, June 2005.

Abbreviations

3G	Third Generation
AAS	Adaptive Antenna Systems
ADC	Analog to Digital Converter
ADSL	Asymmetrical Digital Subscriber Line
AFD	Average Fade Duration
AGC	Automatic Gain Control
AP	Access Point
API	Application Program Interface
ARCS	Astra Return Channel System
ARM	Advanced RISC Machine (processor)
ARQ	Automatic Repeat reQuest
ATM	Asynchronous Transfer Mode
AWGN	Additive White Gaussian Noise propagation channel
BCh	Broadcast Channel
BE	Best Effort Service
BER	Bit Error Rate
BFWA	Broadband Fixed Wireless Access
BRAN	Broadband Radio Access Networks
BS	Base Station
BTS	Base Transceiver System
CATV	Cable Television (originally Community Antenna Television)
CCK	Complementary Code Keying
CID	Connection Identifier
CMOS	Complimentary Metal Oxide Semiconductor
CP	Cyclic Prefix
CPE	Common Phase Error
CPE	Customer Premises Equipment
CRC	Cyclic Redundancy Check
CSI	Channel State Information
CT-2	Cordless Telephone 2nd generation

DAC	Digital to Analog Converter
DAMA	Demand Assignment Multiple Access
DECT	Digital European Cordless Telephony
DFE	Decision-Feedback Equalizer
DFS	Dynamic Frequency Selection
DHCP	Dynamic Host Configuration Protocol
DIUC	Downlink Interval Usage Code
DL	Downlink
DLC	Data Link Control
DMA	Direct Memory Access
DMT	Discrete MultiTone
DOCSIS	Data over Cable System Interface Specification
DSL	Digital Subscriber Line
DSP	Digital Signal Processing
DSSS	Direct Sequence Spread Spectrum
DVB	Digital Video Broadcasting
EC	Echo Canceling
ETSI	European Telecommunication Standards Institute
FCC	Federal Commission on Communications
FDD	Frequency Division Duplex
FEC	Forward Error Correction
FFT	Fast Fourier Transform
FIFO	First-In-First-Out (memory)
FIR	Finite Impulse Response (Filter)
FNC	Foreign Noise Contribution
FSO	Free Space Optics
FTP	File Transfer Protocol
GEO	Geostationary Earth Orbit
GPRS	General Packet Radio Service
GSM	Global System for Mobile communications(original Groupe Spéciale Mobile)
HAP	High Altitude Platform
HFC	Hybrid Fiber Coax
HFFD	Half Frequency Domain Duplexing
HIPERACCESS	High-Performance Access
HIPERLAN	High-Performance Local Area Networks
HIPERMAN	High-Performance Metropolitan Area Networks
HSDPA	High-Speed Downlink Packet Access
HW	Hardware
ICI	Inter Carrier Interference
ID	Identifier
IEEE	Institute of Electrical and Electronics Engineers
IF	Intermediate Frequency
IFFT	Inverse Fast Fourier Transform
iid	Independent Identically Distributed
IL	Implementation Loss
IMEC	Interuniversity MicroElectronics Center
IMT	International Mobile Telecommunications

IP	Internet Protocol
IPv4	Internet Protocol, version 4
IPv6	Internet Protocol, version 6
ISDN	Integrated Services Digital Network
ISI	Inter-Symbol Interference
IWT	Institute for the Promotion of Innovation by Science and Technology in Flanders
Kbps	Kilobits per second
LAN	Local Area Network
LCR	Level Crossing Rate
LEO	Low Earth Orbit
LMDS	Local Multipoint Distribution Service
LMMSE	Linear Minimum Mean-Squared Error
LMS	Least Mean Squares
LOS	Line-of-Sight
LS	Least Squares
LSB	Least Significant Bit
MAC	Medium Access Control
MACh	Multiple Access Channel
Mbps	Megabits per second
MBWA	Mobile Broadband Wireless Access
MCBS-CDMA	Multi-Carrier Block-Spread Code-Division Multiple Access
MC-CDMA	Multi-Carrier Code-Division Multiple Access
Mcps	Megachips per second
MIB	Management Information Base
MIMO	Multi-Input Multi-Output
ML	Maximum likelihood
MLSE	Maximum Likelihood Sequence Estimation
MSB	Most Significant Bit
MMSE	Minimum Mean Square Error
MT	Mobile Terminal
NLOS	Non-Line-of-Sight
nRT-pS	Non Real-Time Polling Service
OECD	Organization for Economic Cooperation and Development
OFDM	Orthogonal Frequency Division Multiplexing
OFDMA	Orthogonal Frequency Division Multiple Access
ONC	Own Noise Contribution
ONU	Optical Network Unit
OSI	Open System Interconnection
PA	Power Amplifier
PAPR	Peak to Average Power Ratio
PC	Personal Computer
PDF	Probability Distribution Function
PDU	Protocol Data Unit
PER	Packet Error Rate
PHS	Personal Handy-phone System
PHS	Payload Header Suppression
PHSF	Payload Header Suppression Field

PHSM	Payload Header Suppression Mask
PHY	Physical layer
PL	Path Loss
PLC	Power Line Communication
PON	Passive Optical Network
POTS	Plain Old Telephone Systems
QAM	Quadrature Amplitude Modulation
QoS	Quality of Service
QPSK	Quadrature Phase Shift Keying
RF	Radio Frequency
RISC	Reduced Instruction Set Computer
RMS	Root Mean Square
RP	Repeater
RRC	Root Raised Cosine
RS	Reed-Solomon (block error code)
RSSI	Received Signal Strength Indicator
RTOS	Real-Time Operating System
RT-pS	Real-Time Polling Service
RX	Reception
RU	Ramp Up
SAP	Service Access Point
SCBS-CDMA	Single-Carrier Block-Spread Code-Division Multiple Access
SC-CDMA	Single-Carrier Code-Division Multiple Access
SC	Single Carrier
SCa	Single Carrier (802.16 variant below 11 GHz)
SDMA	Space-Division Multiple Access
SDU	Service Data Unit
SINR	Signal-to-Interference-and-Noise Ratio
SIR	Signal-to-Interference Ratio
SISO	Single-Input Single-Output
SM	Spatial Multiplexing
SME	Small and Medium Enterprise
SNMP	Simple Network Management Protocol
SNR	Signal-to-Noise Ratio
SS	Subscriber station
STBC	Space-Time Block Coding
STC	Space-Time Coding
STTC	Space-Time Trellis Coding
SUI	Stanford University Interim
SW	Software
TCO	Total Cost of Ownership
TCP	Transmission Control Protocol
TD-CDMA	Time-Division/Code-Division Multiple Access
TDD	Time Division Duplex
TDM	Time-Division Multiplex
TDMA	Time Division Multiple Access
TH	Tomlinson-Harashima
TS	Training Symbol

TX	Transmission
UGS	Unsollicited Grant Service
UL	Uplink
UIUC	Uplink Interval Usage Code
US	United States (of America)
UW	Unique Word
VDSL	Very high speed Digital Subscriber Line
VLAN	Virtual Local Area Network
VSA	Vector Signal Analyzer
WCDMA	Wideband Code-Division Multiple Access
WEP	Wireless Equivalent Privacy
WiFi	Wireless Fidelity
WLAN	Wireless Local Area Network
WLL	Wireless Local Loop
ZF	Zero-Forcing

Variables

a	empirical parameter in propagation exponent for BFWA
a_l	real gain of the lth channel tap
$A_l(t)$	complex gain of the lth channel tap
b	wordlength
b	empirical parameter in propagation exponent for BFWA
B_{aa}	pointing range of an adaptive antenna
B_{ae}	beamwidth of an antenna element
B_{coh}	coherence bandwidth of the channel
B_{mp}	overlap of beamwidth between neighbouring antenna elements
BW	physical bandwidth
c	speed of light
c	empirical parameter in propagation exponent for BFWA
$\mathbf{c}[k]$	M_T-dimensional spatial code vector
$\mathbf{C}$	$M_T \times N_F$ spatial code matrix
d	distance between transmitter and receiver
d_0	reference distance between transmitter and receiver
D_{coh}	coherence distance of the channel
E_{MAC}	efficiency of the MAC
$E_{MAC,0}$	efficiency of the MAC when PER equals 0
$\mathbf{E}$	$M_T \times M_T$ error covariance matrix
f	carrier frequency
$\mathbf{F}$	FFT matrix
f_c	carrier frequency
f_D	Doppler spread or maximum Doppler shift
F_s	sampling frequency
$\mathbf{F}[n]$	$M_T \times M_T$ precoding matrix on the nth subcarrier (frequency domain)
$\mathbf{G}[n]$	$M_R \times M_R$ decoding matrix on the nth subcarrier (frequency domain)
h	CPE antenna height

h_b	BTS antenna height
$h(t)$	complex channel impulse response
$\mathbf{h}$	channel taps
$\mathbf{h_{LMMSE}}$	linear minimum mean squared error estimation of channel taps
$\mathbf{h_{LS}}$	least squares estimation of channel taps
$\mathbf{h_{ML}}$	maximum likelihood estimation of channel taps
$H(f,t)$	channel frequency response at frequency f and observation time t
$\mathbf{H}(\tau,t)$	$M_R \times M_T$ MIMO channel impulse response
$\mathbf{H}_l$	$M_R \times M_T$ complex channel gain of the lth channel tap
$\mathbf{H}_l^f$	$M_R \times M_T$ constant LOS matrix of the lth channel tap
$\mathbf{H}_l^v$	$M_R \times M_T$ complex Gaussian-distributed NLOS matrix of the lth channel tap
$\mathbf{H}[n]$	$M_R \times M_T$ flat fading MIMO channel matrix on the nth subcarrier (frequency domain)
I	In-phase component of a signal
K_l	Ricean K factor of the lth channel tap
l	tap in a mutipath impulse response.
L_p	packet size.
L	number of OFDM symbols between every midamble
$\overline{L}$	average communication distance
M_T	number of transmit antennas
M_R	number of receive antennas
n	discrete time index
$n(t)$	noise signal
N	number of nodes (subscriber stations) within one square meter
N_a	number of antenna elements
N_{BWreq}	number of BW request slots
N_c	number of subcarriers
N_{CBPS}	number of coded bits per OFDM symbol
N_{cp}	length of the cyclic prefix (in samples)
N_d	effective number of data bits in a frame
N_F	number of data symbols per frame
N_h	number of channel taps
N_{hops}	number of hops to communicate from a subscriber station to a base station
N_{oc}	number of open connections
N_{OFDM}	number of OFDM symbols per frame that are available for data transmission
$N_{overheadbits}$	number of overhead bits per frame
$N_{overheadsymbols}$	number of overhead OFDM symbols per frame
N_p	number of symbols in a pilot block
N_u	number of active users (of the MAC)
N'	number of separations between bursts
$\mathbf{n}[n]$	M_R-dimensional received noise vector on the nth subcarrier (frequency domain)

$P(\tau)$	power delay profile
P_{fading}	probability of fading
$P_{h,l}(\nu)$	Doppler power spectrum of the lth channel tap
$P_h(\theta)$	Angular power spectrum
$PL(d)$	path loss for a distance d
$PL_s(d)$	free space path loss for a distance d
r	received signal (time domain)
$\hat{\mathbf{r}}$	estimation of the received signal (time domain)
$\mathbf{r_{iq}}$	received signal distorted by I/Q mismatch (time domain)
$\mathbf{r_l}$	received signal for the lth OFDM symbol (time domain)
$\mathbf{r_p}$	received signal for the pilot symbol (time domain)
R	effective data rate per node (subscriber station)
R_{mode}	physical communication rate for a communication mode
$\mathbf{R}_{Tx}^{l}$	transmit spatial covariance matrix of the lth channel tap
$\mathbf{R}_{Rx}^{l}$	receive spatial covariance matrix of the lth channel tap
$R(\Delta d)$	spaced-distance correlation function over Δd
$R_A(\Delta f)$	spaced-frequency correlation function of the channel frequency response over Δf
$R_l(\Delta t)$	spaced-time correlation function of the lth channel tap over Δt
s	shadowing term in path loss
s	transmitted data symbol
$\hat{\mathbf{s}}$	estimate (at receiver) of transmitted data symbol
$s[k]$	k^{th} input data symbol
$\mathbf{s}[n]$	M_T-dimensional transmitted symbol vector on the nth subcarrier (frequency domain)
$\mathbf{s_p}$	transmitted data symbol vector for the OFDM pilot symbol (frequency domain)
t	observation time
t	frequency-domain transmitted symbol vector
$\overline{T}_{burst}$	average burst duration
T_b	bit duration
T_{coh}	coherence time of the channel
T_f	frame duration
T_{OFDM}	OFDM symbol duration
T_s	symbol duration
Q	quadrature component of a signal
ν	receiver velocity
W	throughput per subscriber station
$x(n)$	signal (discrete time)
$\mathbf{x_p}$	transmitted data symbol vector for the OFDM pilot symbol (time domain)
y	received frequency domain symbol vector .
$\mathbf{y}[n]$	M_R-dimensional received vector on the nth subcarrier (frequency domain)

y_{iq}	received frequency domain symbol vector distorted by I/Q mismatch.
y_m	mirrored version of the received frequency domain symbol vector .
y_p	received data symbol vector for the OFDM pilot symbol (frequency domain)
α	attenuation coefficient
α	roll-off factor of root-raised cosine filter
α	I/Q mismatch distortion coeeficient
β	minimum signal-to-noise ratio for successful reception
β	I/Q mismatch distortion coeeficient
$\delta(t)$	delta function
Δd	distance separation
Δf	frequency separation
Δf_c	carrier frequency offset
$\hat{\Delta f}_c$	estimation of the carrier frequency offset
Δn	discrete time separation (in symbols)
ΔPL_f	frequency correction term on path loss
ΔPL_h	CPE antenna height correction term on path loss
Δt	time separation
ΔT_i	time separation between burst i and $i+1$
$\overline{\Delta T}_{bursts}$	average separation between two consecutive bursts
ε_{ML}	Maximum likelihood estimation of symbol timing
ϕ	grazing angle of internal wall
γ	path loss exponent
λ	wavelength
μ	normalized clipping level
ν	Doppler frequency variable
θ	grazing angle of external wall
$\overline{\theta}$	mean arrival angle
θ_l	phase shift of the lth channel tap
$\rho(\Delta t)$	correlation coefficient for amplitude fading at Δt
$\rho_{i_1 i_2}^{Tx,l}$	spatial correlation between transmit antennas i_1 and i_2 of the lth channel tap
$\rho_{j_1 j_2}^{Rx,l}$	spatial correlation between receive antennas j_1 and j_2 of the lth channel tap
σ_θ	rms angle spread
$\overline{\tau}$	average excess delay of a channel
τ	delay time
τ_l	delay of the lth channel tap
τ_{max}	maximum excess delay of a channel
τ_{RMS}	RMS delay spread of a channel

Notation

$\mathbf{x}$	vector $\mathbf{x}$
$\mathbf{X}$	matrix $\mathbf{X}$
$\mathbf{X}^{\mathrm{T}}$	transpose of matrix $\mathbf{X}$
$\mathbf{X}^{\mathrm{H}}$	Hermitian transpose of matrix $\mathbf{X}$
$\mathbf{X}^{*}$	complex conjugate of matrix $\mathbf{X}$
$\mathbf{X}^{-1}$	inverse of matrix $\mathbf{X}$
$\mathrm{pinv}(\mathbf{X})$	Moore-Penrose pseudo-inverse of matrix $\mathbf{X}$
$\|\mathbf{X}\|$	Frobenius norm of matrix $\mathbf{X}$
$\mathrm{diag}(\mathbf{x})$	square diagonal matrix with the elements of vector $\mathbf{x}$ as diagonal
$E[x(t)]$	Expected value of $x(t)$
$\mathrm{var}[x(t)]$	variance of $x(t)$
$\arg\max_{i}(f(i))$	returns i for which $f(i)$ is maximized
$\tan^{-1}(x)$	inverse tangent of x

Index

SIGNALS AND COMMUNICATION TECHNOLOGY

(continued from page ii)

Chaos-Based Digital Communication Systems
Operating Principles, Analysis Methods, and
Performance Evalutation
F.C.M. Lau and C.K. Tse
ISBN 3-540-00602-8

Adaptive Signal Processing
Application to Real-World Problems
J. Benesty and Y. Huang (Eds.)
ISBN 3-540-00051-8

**Multimedia Information Retrieval and
Management Technological**
Fundamentals and Applications D. Feng, W.C.
Siu, and H.J. Zhang (Eds.)
ISBN 3-540-00244-8

Structured Cable Systems
A.B. Semenov, S.K. Strizhakov,and I.R.
Suncheley
ISBN 3-540-43000-8

UMTS
The Physical Layer of the Universal Mobilc
Telecommunications System
A. Springer and R. Weigel
ISBN 3-540-42162-9

Advanced Theory of Signal Detection
Weak Signal Detection in Generalized
Obeservations
I. Song, J. Bae, and S.Y. Kim
ISBN 3-540-43064-4

Wireless Internet Access over GSMand UMTS
M. Taferner and E. Bonek
ISBN 3-540-42551-9